UNITED STATES
DEPARTMENT OF THE INTERIOR
FISH AND WILDLIFE SERVICE

EASTERN ENERGY AND LAND USE TEAM
Route 3, Box 44
Kearneysville, West Virginia 25430

Dear Colleague:

We are pleased to provide you with a copy of "Planning for Urban Fishing and Waterfront Recreation" FWS/OBS - 80/35 a reference manual for assisting administrators and planners with decisions related to incorporating fishing and related recreation into urban waterfront developments. This document came about largely through the efforts of Mr. Carl Sullivan, Executive Director, American Fisheries Society, Bethesda, Maryland, who along with officials of other conservation groups, saw the many public benefits of an easy-to-use reference manual on the subject, quite similar to the highly successful "Planning for Wildlife in Cities and Suburbs" FWS/OBS - 77/66 (January 1980).

The authors of "Planning for Urban Fishing and Waterfront Recreation" review past national experiences on the subject and tie in case examples, provide a brief overview on aquatic life processes, and present a step-wise planning process for gathering information and assistance to incorporate fishing and recreation into waterfront developments. The Appendices contain abundant information sources for use by local decisionmakers.

Your evaluation of this document will help the Team and the Office of Biological Services with future project planning. Please address comments to the Team Leader.

Sincerely,

Edgar A. Pash

Edgar A. Pash
Team Leader, EELUT

For sale by the Superintendent of Documents, U.S. Government Printing Office
Washington, D.C. 20402

FWS/OBS-80/35
July 1981

Planning for Urban Fishing and Waterfront Recreation

by

D. L. Leedy
T. M. Franklin
R. M. Maestro

Urban Wildlife Research Center, Inc.
10921 Trotting Ridge Way
Columbia, Maryland 21044

Fish and Wildlife Service
Cooperative Agreement Number 14-16-0009-79-1014

William T. Mason, Jr., Project Officer

Eastern Energy and Land Use Team
U.S. Fish and Wildlife Service
Route 3, Box 44
Kearneysville, West Virginia 25430

In cooperation with

Heritage Conservation and Recreation Service
U.S. Department of the Interior
Washington, DC

William Honore, Project Officer

Water and Power Resources Service

Lawrence Roberts, Project Officer

Water Resources Institute
U.S. Army Corps of Engineers
Washington, DC

Mary K. Vincent, Project Officer

Clean Lakes Program
U.S. Environmental Protection Agency
Washington, DC

Michael Moyer, Project Officer

Cosponsored by

Sport Fishing Institute, Washington, DC

Robert Martin, Project Officer

The Academy of Natural Sciences of Philadelphia
Philadelphia, Pennsylvania 19103

Lynn Berseth, Project Officer

Performed for

Office of Biological Services
Habitat Preservation Program
Fish and Wildlife Service
U.S. Department of the Interior
Washington, DC 20240

Preface

Urban developers are beginning to realize that fish and wildlife are integral parts of the environment and that a healthy environment can contribute to social, economic, and aesthetic improvement. Likewise, planners are learning the importance of incorporating fish and wildlife provisions into their urban designs and are looking for guidelines and information to help them.

This guide complements one entitled *Planning for Wildlife in Cities and Suburbs,** which described the incorporation of wildlife considerations in urban development and redevelopment projects. Readers of this work wanted a similar volume emphasizing ways by which fish and other aquatic resources could be incorporated into urban area plans for man's enjoyment.

Urban and regional plannners, developers, recreation and water resource planners, politicians, and public and private organizations concerned with urban planning and management should find this document useful. By reading it, they will gain an appreciation of, and practicable advice on how to incorporate fish, wildlife, and recreational values into their decision-making and planning procedures. Planners having little knowledge about fish, wildlife, or ecology may refer to this guide whenever they require reliable information about reasoned approaches and methods to use in their work.

Contributions by biologists, ecologists, and other specialists not only can help planners and developers to protect or provide habitats suitable to fish and other aquatic life, but can also aid them in managing waterfronts so that their recreational use and enjoyment by urban residents will be enhanced. Information sources and actual cases showing successful fish, wildlife, and recreation value integration into the planning and management of urban and suburban waterfronts are given.

Hopefully, this guide will serve the following functions:

- prod planners and developers concerning what they can do to promote aquatic resources conservation in urban and urbanizing areas, and provide useful information for doing so;
- alert and inform public officials at state and municipal levels about opportunities for cooperative development of urban waterfronts using federal and other types of technical and financial assistance;
- encourage planners, developers, biologists, engineers, and the public to plan and manage together;
- stimulate environmental activists and civic organizations to engage in projects facilitating the preservation, improvement, or development of new, urban aquatic habitats; and
- provide universities, schools, and other organizations with helpful materials and ideas in training, research, public education, and interpretive programs.

*Available from Urban Wildlife Research Center, 10921 Trotting Ridge Way, Columbia, Maryland 21044; The American Planning Association, 1313 East 60th Street, Chicago, Illinois 60637, and the U.S. Government Printing Office, Washington, DC.

The reader should be aware that program adjustments have affected some of the cosponsors of this document. The Heritage Conservation and Recreation Service, U.S. Department of the Interior, was abolished on May 31, 1981. The program will merge into the National Park Service. The President's FY 82 budget did not recommend funding for new cooperative agreements for the U.S. Environmental Protection Agency's Clean Lakes Program. As of May 18th, 1981, the Secretary of the Interior changed the name of the Water and Power Resources Service to the Bureau of Reclamation. The reader should be aware that statements relative to the Water and Power Resources Service should be applied to the Bureau of Reclamation. Questions pertaining to statements in this document that concern the above offices should be directed to the appropriate Department/Agency.

Executive Summary

To help planners and developers in enhancing urban fishing and waterfront recreation, information about aquatic resources and urbanization effects on those resources is presented. The central theme revolves around opportunities for (a) preserving existing high-quality aquatic areas, (b) restoring degraded areas, and (c) creating new areas where appropriate.

Part I gives background information about environmental, fish, and wildlife values, the nature of aquatic ecosystems, urbanization effects on these systems, planning and management implications, and the importance of aquatic resource considerations in urban-suburban planning.

Part II deals specifically with planning for urban fishing and related waterfront recreation. Literature on past and current urban fishing programs is reviewed for information on requirements, opportunities, applicable laws, and constraints. For those interested in promoting urban fishing, progressive guidance steps are included, as are suggestions for conserving high-quality fishing waters, improving deteriorated waters, and new water body creation. Increasing urban fishing opportunities and enjoyment by providing access and facilities for anglers, stocking, fishing in currently unused water bodies, installing fish attractors, and other means are also addressed. Similar treatment is given to the enhancement of other forms of water-based or waterfront recreation including swimming, boating, canoeing, water skiing, nature watching, pleasure driving, biking, picnicking, and hiking along the waterfront. Attention is paid to zoning, liability, safety, and other considerations.

Part III provides detailed guidance on steps that can be taken by planners to incorporate considerations for aquatic resources into the planning and decision-making process to ensure that healthy water bodies will be retained, degraded waters will be rehabilitated, and new water bodies will be created for specific or multipurpose uses. Chapter 6 describes steps for site planning and Chapter 7 outlines steps having special consideration for regional and municipal planning, including a checklist of items for review before implementing the aquatic resources plan.

Though numerous references are cited in footnotes throughout the guide, Part IV provides additional sources of technical information and financial assistance. Funding opportunities and requirements under various federal laws are described. A chapter is also included on recommended readings.

Appendices to the guide include names and addresses of various federal, regional, and state agencies, together with tables useful to planners and developers as references and guides for making aquatic resource inventories and understanding water quality parameters.

Table of Contents

Acknowledgements

The authors and sponsors gratefully acknowledge the many significant contributions made by the individuals named here who have materially assisted in the development of the content and readability of this report.

Carl R. Sullivan, Executive Director of the American Fisheries Society, provided much of the impetus for the preparation of this guidebook in a letter to Lynn A. Greenwalt, Director of the U.S. Fish and Wildlife Service. Project feasibility was explored with Sullivan, Richard H. Stroud, and Robert G. Martin of the Sport Fishing Institute, Raymond E. Johnson, former president of the American Fisheries Society, and representatives of the Fish and Wildlife Service. Interest grew as discussions were held with representatives of other agencies who expressed the need for a broadened scope, including not only urban fishing but other water-related outdoor recreation and waterfront activities requiring healthy aquatic habitats in cities and suburbs.

Ann Breen Cowey, Office of Coastal Zone Management, National Oceanic and Atmospheric Administration, and Herbert B. Quinn and Harvey Doerksen, Office of Biological Services, Washington, DC, offered many valuable suggestions on content and organization.

Careful manuscript reviews by Kenneth M. Mackenthun, Vice President, Enwright Laboratories, Inc., are acknowledged with thanks as are comments by Thomas Dolan, natural resource consultant and member of the Board of Directors, Urban Wildlife Research Center. Several others to whom the guidebook or parts of it were sent, provided helpful information or comments. Among these were: Keith A. Argow, Executive Director, and Pamela K. McClelland, Trout Unlimited; Thomas N. Shiflet, Director, Ecological Sciences and Technology Division; Warren J. Fitzgerald, Illinois State Conservationist; Gerald R. Calhoun, Maryland State Conservationist; and Hans G. Uhlig, Head, Conservation Technology Staff, Northeast Technical Service Center, Soil Conservation Service; J. H. Coombes, Chief, Engineering Division, and Mark Sifuentes, Outdoor Recreation Planner, Albuquerque District Corps of Engineers; Steven R. Alcorn, Fisheries Management Biologist, Missouri Department of Conservation; Robert F. Hutton, Recreational Fisheries Coordinator, National Marine Fisheries Service, NOAA; and Raymond E. Johnson. Additional comments and information were provided by several of the Heritage Conservation and Recreation Service Regional Offices and by Willard T. Johns, Director, Office of Information, Pennsylvania Fish Commission.

Special thanks are tendered to Charles Thurow, Assistant Director of Research of the American Planning Association, for his thoughtful review and comment on various drafts of the guide.

Bryan L. Swift, wildlife biologist, National Water Resources Analysis Group, and Robert A. Honig, Ecological Intern, Institute of Ecology, Indianapolis, Indiana, working with the OBS Eastern Energy and Land Use Team, Kearneysville, West Virginia, made many useful suggestions. Within the Urban Wildlife Research Center, Louise Dove, Research Assistant, obtained many of the references used in preparing the guide, and Joyce Keifer and Barbara McFalls typed several drafts of the manuscript.

The Academy of Natural Sciences of Philadelphia, as a subcontractor, provided much of the information on which Chapters 2, 6, and 7 of the guide are based. This information was developed primarily by Lynne F. Berseth and Paul E. Stacey of the Academy's Division of Limnology and Ecology. Responsibility for selecting the material used and for its presentation is assumed by the authors.

Part I

Background Information

Chapter 1

Introduction

Purpose and Intended Users

Municipal, regional, urban, recreation, and water resources planners, developers, and officials concerned with land use planning, urban development, and fish and wildlife management will find this guide useful. Others who may find it a valuable reference work include landscape architects, engineers, planning boards, civic organizations, aquatic biologists, professors, students, environmentalists, and private citizens.

Scope and Definitions

The scope of the guide is nationwide, but regional constraints or differences are noted when pertinent. The term "urban-suburban areas" is used in a broad sense to include towns, villages, cities, metropolitan areas, and aquatic resources near these areas. "Urbanizing areas" are those being developed, whether for housing, airports, streets, industry, or sewage treatment plants. "Aquatic resources" include any surface waters, ranging from streams, rivers,

Photo: D. Chavalier, *The Washington Post*

Figure 1. *Fishing provides high quality recreation for many urban residents.*

Figure 2. *Pennypack Park, part of Philadelphia's Fairmount Park System, leaves a narrow strip of open park land amid large areas of urban expansion.*

and lakes to estuarine and coastal waters, manmade ponds, water supply reservoirs, wet gravel pits, sewage lagoons, etc., together with the aquatic organisms associated with these waters. Also, examples include other water-dependent animals, such as birds, amphibians, reptiles, and mammals in urban waters and suggestions for their management.

This guide provides background information on aquatic resources and the need for their consideration in urban planning and management (Part I), describes planning approaches for urban fishing and other waterfront recreation (Part II), suggests how aquatic resource considerations can be integrated into planning and decision-making so that they are available for recreation or other uses following urbanization (Part III), and provides sources of technical and financial assistance for urban-suburban planning and development (Part IV). Brief discussions of fish and wildlife management principles and approaches are included to alert planners and developers to ways in which their decisions and actions can enhance or diminish urban fishing and other waterfront recreation opportunities. Readers may refer to Part IV and Recommended Readings for more detailed information.

The Setting

About three of every four persons in the United States live in cities, towns, or suburbs. Many people live in the heart of metropolitan areas where there is little opportunity to fish or enjoy nature. Today's high transportation costs prevent inner city residents from traveling to rural areas, so city parks having access to fishable waters, riverside walkways, waterfront facilities, and natural recreation areas become increasingly desirable.

Seventy percent of the 415 cities in the United States having a population of 50,000, and at least 30 percent of the 520 cities having a population between 25,000 and 50,000, are located on the edge of a river, lake, bay, or ocean.[1] It is estimated that over 54 percent of the nation's population, excluding Alaska and Hawaii, lives within 50 miles of the coastal zone.[2] It would seem that with this distribution of human population in relation to water

[1]USDI Heritage Conservation and Recreation Service, *Urban Waterfront Revitalization: The Role of Recreation and Heritage, Vol. 1, Key Factors, Needs and Goals,* undated, 31 pp.

[2]D. Wren, "Beach Access," Page 3 in *Environmental Comment,* 1980. (Urban Land Institute, Washington, DC)

bodies, there would be sufficient water for urban fishing and other water-based recreation; however, this is not the case. One reason for this is the extent to which aquatic habitats have been degraded through urbanization, industrialization, and other land and water uses. Fortunately, some waters have been cleaned up and others can be. An estimated 80 percent of more than 3,700 urban lakes in the United States are significantly degraded, and yet they offer potential aesthetic and recreational value to more than 94 million metropolitan residents.[3]

Why Consider Aquatic Resources in Urban-Suburban Planning?

There are many reasons why water, and the plants and animals depending on it should be considered in urban-suburban planning. Aquatic resources are an integral part of the environment, and water, essential to life, is needed for a variety of domestic and industrial uses. Furthermore, planners, developers, and residents alike recognize that a high-quality environment in which to live, work, or play, has many values. Though difficult to measure, the aesthetic value of an attractive, safe river or lake in an urban setting is important. Other reasons include:

- Fish and wildlife share the earth with us and are an inherent component of the environment.
- Urban-suburban residents enjoy seeing wildlife in the vicinity of their homes and many like to fish or engage in other water-based recreation close to where they live.
- Anything affecting the environment—soil, water, vegetation, and air—affects fish and wildlife and, eventually, humans, who have more control over the environment than any other species.
- Generally speaking, water and land conditions favorable for desirable fish and wildlife are also favorable for man. On the contrary, waters unsuitable for game fish because of pollution, nuisance plant growth, or other characteristics do not have the qualities desired for water-based recreation activities like swimming or water skiing.
- Construction of well designed and suitably located recreation lakes, fishing piers, marinas, boat launching sites, water-side parking, picnicking, and sanitation facilities can add to the attractiveness of an urban area, and increase real estate values. Greater recreational use also results in expenditure for supplies and equipment in local stores and elsewhere, thus stimulating the economy.
- Maintaining natural sites in or near metropolitan areas and encouraging visits by elementary school classes facilitates environmental education which can create public support for conservation programs both within and outside of urban areas.[4] Urban natural areas have educational value for adults directly, too. Using a closed laundry and a previously polluted mill pond in New Hampshire as a "Valley Natural Area," Louis and Margery Milne[5] demonstrated how aquatic wildlife can be used to interest and convince adult citizens that they *can* do things to enrich the environment.
- Planners and developers are required by various federal, state, and local regulations to consider the environmental impacts of development on fish and wildlife, particularly species that are threatened or endangered.
- Finally, the planning and development of urban and suburban communities to optimize positive, beneficial impacts on fish and wildlife, while minimizing negative effects, represents a challenge to all concerned.

What Planners and Developers Can Do

Planners and developers, with technical advice from aquatic ecologists and other professionals can do much to provide the environment needed for fish and wildlife incorporation into urban and suburban areas. They can set the stage for future fish and wildlife populations and their management in these areas by protecting unique biological communities, maintaining free-flowing streams having vegetated borders rather than channelizing them or enclosing them in concrete conduits, preserving wetlands rather than draining them, designing storm water control ponds and sediment basins to facilitate their cleaning and maintenance so that fish and wildlife can use them after construction, encouraging cluster housing for more open space, including wetlands, and creating productive aquatic habitats where none exists. Urban and suburban planners and developers should become fish and wildlife planners and managers, too, a goal that will become clearer in succeeding chapters.

[3]Council on Environmental Quality, *Environmental Quality: the Tenth Annual Report of the Council on Environmental Quality*, 1979. 816 pp. (Executive Office of the President, Council on Environmental Quality, 722 Jackson Place, N.W., Washington, DC 20006.)

[4]R. A. MacMullan, *Meeting Urban Wildlife Needs*, pp. 31-37 in Man and Nature in the City, proceedings of a symposium, 1968. (U.S. Department of the Interior.)

[5]Louis and Margery Milne, *Urban Wildlife as a Tool in Education*, pp. 167-169 in the proceedings of a symposium, Wildlife in an Urbanizing Environment, 1974. (University of Massachusetts.)

Chapter 2

The Nature of Aquatic Ecosystems

This chapter provides an abbreviated account about the nature and functioning of aquatic ecosystems. It serves as background information to help planners understand and use the planning approaches suggested later in this guide.

An ecosystem may be defined as an ecological community considered together with the nonliving factors of its environment as a unit. One of these nonliving factors—water—plays a critical role in the survival of all life forms and, for many species, is the medium in which they live all or part of their lives. Surface water, with which we are primarily concerned, exists in a variety of locations and forms ranging from tiny springs to mighty rivers and from small ponds to huge lakes. Wetlands, whether they form large, slowly-flowing marshes or are isolated in lowlands, provide habitats for many aquatic and terrestrial life forms. Of great importance, also, are estuarine and coastal waters. Life in some form exists in essentially all of these waters. Regardless of water type, certain basic, ecological principles apply to all aquatic ecosystems.

Energy Cycling

All organisms depend on, and are involved in, energy cycling. A simple model of energy flow (Figure 3) places primary producers (plants, algae) at the lowest trophic (nutritional) level. These organisms use nutrients in the water and energy from the sun to produce a living biomass (photosynthesis). The next trophic level is composed of herbivores or vegetation-eaters which graze on the plants or filter algae and bacteria from the water to create their biomass. In aquatic systems, zooplankton (microscopic invertebrates), insects, and a few fish may be considered herbivores. At the top of the food chain are the carnivores, including most fish, which rely on animal biomass to meet their energy requirements. Solar energy passes from the sun to the primary producers, to the herbivores, to the carnivores and occasionally to secondary carnivores in a simple model. Ending the chain are decomposers which convert dead plant and animal matter to nutrients, which again feed the plants. These decomposers also serve an integral function throughout the energy chain. Some biomass is lost to the decomposers along the way in the form of plants or animals which die before being consumed by members of the next trophic level. Some is lost in the form of metabolic wastes from animals. This simple model is diagrammed in Figure 3.

Aquatic Habitat Attributes

Aquatic organisms have specific habitat requirements which must be met to ensure perpetuation of the species. Physical and chemical interactions making up the aquatic environment are complex. Of the physical attributes, current, temperature, sunlight penetration, substrate, and physical configuration are the most important. Dissolved gases, dissolved solids, salinity, nutrients, and organic matter are all important components of water quality in which aquatic life is found.

Current

In lakes, where the water is relatively still (lentic), in contrast to the flowing (lotic) environment of rivers, free-floating algae and shore-zone (littoral) rooted plants provide most of the plant matter upon which higher trophic levels rely. In rivers where the current is relatively rapid, free-floating algae are quickly washed downstream and are unable to establish themselves. Also, periphyton (attached algae et al.) growing on rocks and other substrates is a very important food source for aquatic animals. In these rivers, much of the plant matter and other detritus is washed into the system by water flowing throughout the watershed, or falls into the rivers from overhanging vegetation. This material finds its way into the energy cycle through processing by specialized insects (detritivores) that feed on it. Thus, current establishes the mode of primary energy input. Herbivores in lakes and ponds are largely zooplankton feeding on microscopic algae (phytoplankton). In streams, detritus-eating species such as stoneflies, mayflies, and caddisflies are the dominant herbivores. Fishermen recognize these as important fish food.

Temperature

All plant and animal species have thermal requirements or tolerances which limit their ability to survive or reproduce in a particular locality. Thus, cold-water fish like trout and salmon may be present in lakes or streams where temperatures do not exceed 70°F. Warm-water species like bass, sunfish, bullheads, and carp, although able to tolerate winter temperatures near freezing, generally do best in waters where summer temperatures range between 60° and 85°F. Plants and invertebrates show similar relationships.

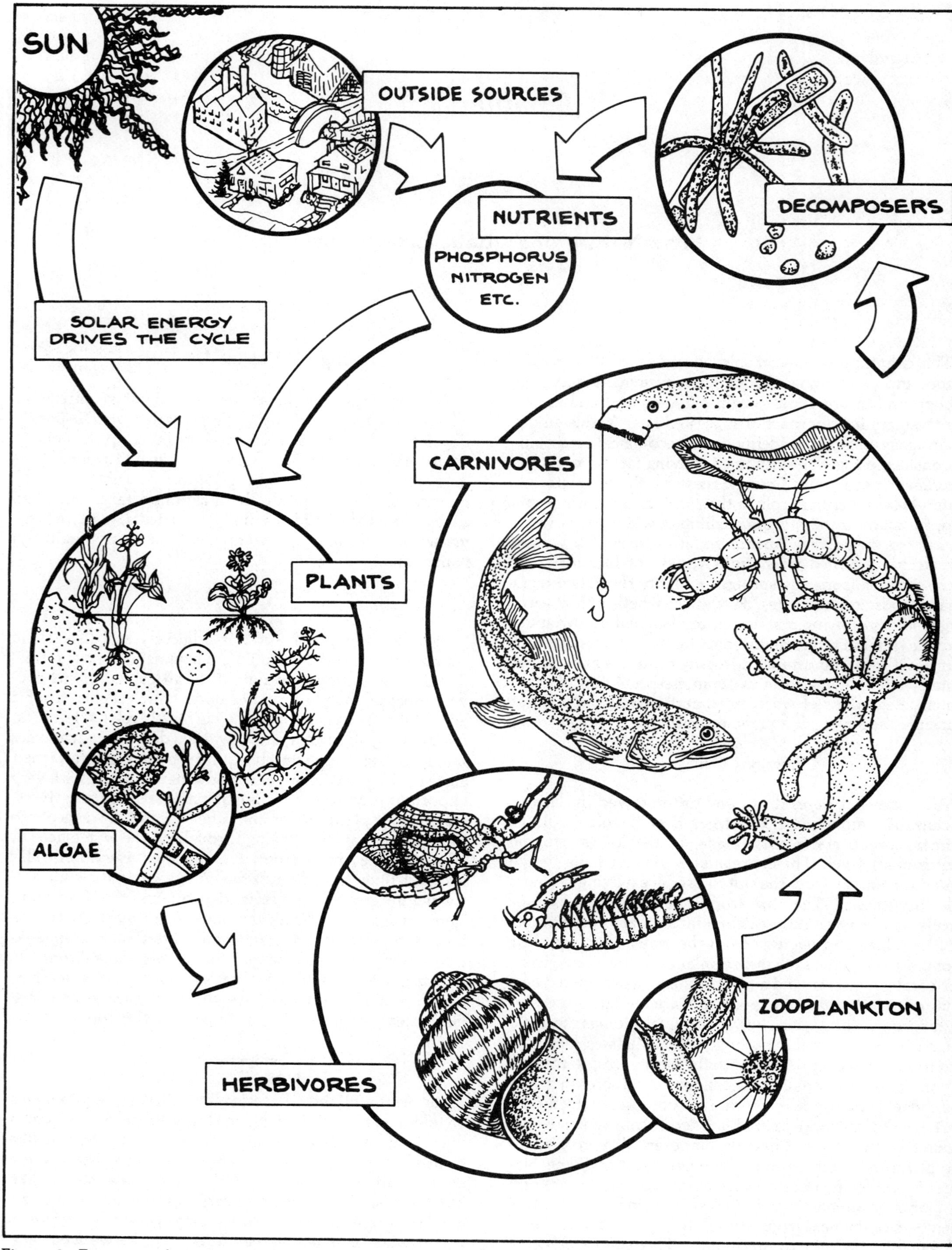

Source: U.S. Environmental Protection Agency, *Clean Lakes and Us.* EPA-440/5-79-021, Office of Planning and Standards, Washington, DC 20460.

Figure 3. *Energy cycle.*

Temperature plays a critical role, too, in the natural history of many temperate lakes where temperature stratification and mixing (fall and spring "turnovers") of surface and bottom waters occur, thus redistributing nutrients and oxygen. Turnovers occur when water reaches its greatest density at about 39°F.

Sunlight Penetration

Sunlight provides energy to continue photosynthesis; the depth to which sunlight penetrates a lake or pond is a determinant of a lake's productivity. A turbid, silty lake may be rich in nutrients, but low in productivity if photosynthetic activity is reduced by the shading effect. However, turbidity may also result from excessive phytoplankton, indicating high productivity. In rivers, sunlight is needed for the limited in-stream (*in situ* or autochthonous) primary production which does occur.

Substrate and Physical Configuration

In general, a water body's bottom or substrate, if composed of diverse material sizes, e.g., silt, sand, gravel, rubble, and boulders, will provide a wider range of habitats and ultimately a more diverse biological community than one having a uniform substrate like bedrock or sand. Different plants and invertebrates require different substrates, as do fish. Sunfish, for example, need a relatively silt-free substrate, while carp can tolerate more silty conditions.

Lakes having the most convoluted shorelines and extensive littoral (shallow, shore) zones are likely to produce the most fish because most fish species rely on these areas for breeding, feeding, or shelter. Rooted plants also growing in these areas provide food and shelter for insects which, in turn, are consumed by fish. Steep-banked lakes do not provide the littoral habitat found in convoluted lakes having shallow water and wetland vegetation (rooted aquatics) around their shores.

Meandering streams are also more productive, since they usually have pools, rapids (riffles), and a variety of substrates. Attached algae (periphyton) grow best on stable rock substrates, and rubble provides the best habitat for many aquatic insects; however, dragonfly naiads and chironomids (midges) may prefer silty backwaters for burrowing. Many species like trout spawn on gravelly riffles, while bass and sunfish need slow water having a firm substrate for nesting. A meandering stream provides more valuable "edge" habitat than most channelized streams, a factor important to fish and wildlife alike. Catfish and bullheads use undercut stream banks for breeding and shelter. The predatory mink also feeds in these same areas.

Dissolved Gases

Dissolved gases, primarily oxygen and carbon dioxide, are needed by most organisms residing in aquatic systems. Oxygen in ponds and lakes derives from the atmosphere, from photosynthesizing phytoplankton, and from other plants. Carbon dioxide used by plants in photosynthesis comes from atmospheric exchange and from animal respiration. Because sunlight does not penetrate to the bottom of deep lakes, especially during winter when ice and snow block out sunlight over long periods, oxygen depletion may occur resulting in the death of fish and other aquatic organisms. This problem may be particularly acute in highly productive lakes as aquatic vegetation dies and decays, thus adding to the oxygen demand during the unlighted period. Streams exhibit fewer problems with dissolved gases because air is freely exchanged when water passes over rapids and falls.

However, slow-moving streams or marshy, estuarine areas may experience localized oxygen depletion if organic decay is excessive and temperatures are high.

Dissolved Solids, Nutrients, and Organic Matter

Dissolved solids, nutrients, and organic matter all contribute to the productivity of aquatic ecosystems. If one nutrient essential to plant growth or reproduction is absent, even if all others are present in abundance, growth will be limited. Nutrients are added to the aquatic system naturally: through weathering of rock and surface runoff; as organic matter from plants broken down by aquatic animals and bacteria; and by leaching of decaying organic matter and soils. Generally, the more dissolved solids, nutrients, and organic matter entering an aquatic system, the higher the productivity. Estuaries are highly productive because they are the final resting area for much of the nutrient load gathered from miles of stream courses draining many square miles of watershed.

Total dissolved solids in a system also determine the water's salinity. Fresh water salinity is less than 0.5 parts per thousand salt compared to 35 parts per thousand in sea water. Inland waters west of the Rocky Mountains are usually more saline than those in the East. Some fish species have a narrow tolerance to salinity, but anadromous fish migrating from salt to fresh water have a wider salt tolerance. A critical salinity balance exists in most estuarine habitats, but floods and droughts may upset it.

Aging and Succession in Aquatic Ecosystems

Natural complexities make each water body different. Yet under similar conditions, similar biological communities obtain. Over the long term, however, water bodies mature and biological community structures change. As lakes age they change from young, relatively unproductive systems (oligotrophic), to mature, highly productive (eutrophic) ones. In eutrophic lakes, nutrients added to the system are retained, allowing higher primary and secondary production. Eventually, a lake may fill with organic matter and sediment, first forming a bog and finally, dry land. These physical and chemical transitions permit changes in community structure, i.e., eutrophic lakes support warm water fish rather than the trout and cisco which can live only in the cool waters of oligotrophic lakes.

Stream characteristics differ from source to mouth throughout their history, i.e., headwaters are cooler and steeper, contain less nutrients, and support a less productive community with cold water organisms like stoneflies and trout predominating. Lower stream portions exhibit a larger discharge, and a sandier and siltier substrate. And they are warmer, contain more nutrients and organic matter, and support species like bass, gar, fly larvae, and rooted plants in the more extensive backwater areas. As streams mature, the most radical changes are spatial or longitudinal.

Chapter 3

Urbanization Effects on Aquatic Resources and Their Implications for Planning and Management

A literature review confirms that urbanization has both damaging and beneficial effects on the fish and wildlife environment. Though much has been written about destructive environmental impacts, there is less documentation on constructive ones. Little attention has been paid to the effects on fish and wildlife of environmental changes caused by urbanization, how habitat degradation can be reduced through planning and management, or habitat improvement and creation methods to use to prevent habitat elimination. These issues are addressed in this and succeeding chapters. Additional information is contained in an annotated bibliography.[6]

Partial Checklist of Urbanization Effects on Aquatic Resources

This checklist, together with examples, relates to aquatic ecosystem attributes discussed in the previous chapter and to urbanization impacts on fish and wildlife. Because different factors interact, some of the fish and wildlife responses to changed habitat conditions cited may reflect responses to several impacts. The checklist is intended to remind planners, decision-makers, and developers about the environmental impacts, or potential impacts, on their plans or activities.

1. Nutrient additions to aquatic systems

Because people are concentrated in cities and suburbs, their food, raw materials, and construction materials are imported. Urban waste disposal is also an important problem.

Sewage treatment plants, though much improved in recent years by legal regulation and federal financial assistance, continue to contribute nutrients such as nitrogen and phosphorus to aquatic systems. Suburban recreation lakes surrounded by dwellings having inefficient septic tank/waste disposal fields receive excessive amounts of such nutrients. Also, nutrients from fertilizers used on lawns, gardens, and golf courses enter urban waters through runoff and storm sewers.

Dissolved gas concentrations are often upset by the addition of sewage and other oxygen-demanding substances. As these substances decay or oxidize, the amount of oxygen available for the natural fauna is limited, causing many species to die. Sludge worms, carp, and other species which tolerate low oxygen conditions may proliferate. In rivers, low oxygen or septic zones may bar fish from upstream migration. Over-enriched waters often support dense growths of algae and other aquatic plants which interfere with recreational use and, upon dying, create unpleasant odors and contribute further to oxygen depletion.

A study of fish in Goose Creek, Jefferson County, Kentucky, a stream under urban development influence, suggested that during early urbanization stages, fish populations may have been enhanced owing to additional nutrients that had entered the stream. However, sedimentation and discharges from sewage treatment plants soon began to affect the fish distribution adversely.[7]

Within London's inner city, the Thames River was described as being little better than an open sewer in the 1940s and 1950s. It was essentially devoid of water life except for specialized forms that could survive under anaerobic (absence of free oxygen) conditions. With the clean-up of the river, fish and birds returned.[8]

After Lake Washington in Seattle, Washington, had received increasing amounts of treated sewage, desirable forms of plankton characterizing the lake between 1933 and 1950 had been largely replaced by undesirable forms causing nuisance blooms by 1955. Following more effective sewage treatment and diversion of sewage effluent from the lake, conditions for aquatic life improved.[9]

[6]Daniel L. Leedy, *An Annotated Bibliography on Planning and Management for Urban-Suburban Wildlife*, U.S. Fish and Wildlife Service publication FWS/OBS-79/25, 1979, 256 pp. (Available from the U.S. Fish and Wildlife Service—Office of Biological Services, the Urban Wildlife Research Center, Inc., and the U.S. Government Printing Office.)

[7]D. S. White, F. C. Hill, and K. H. Haag, *The Fishes of Goose Creek, Jefferson County, Kentucky: a stream under the influence of urban development*, pp. 45-55 in Transactions, Kentucky Academy of Sciences, Vol. 38 (1-2), 1977.

[8]J. Harrison and P. Grant, *The Thames Transformed: London's River and its Waterfowl*, Andre Deutsch Limited, 1976, 240 pp. (Available in U.S. From Transatlantic Arts, Inc., North Village Green, Levittown, NY 11756.)

Robert G. Halstead, USDA, Soil Conservation Service

Figure 4. *Sediment deposits from development construction have almost completely filled in and blocked up this ditch.*

2. Additions of toxic materials and wastes to urban waters

Urban rivers, lakes, and coastal harbors receive other domestic and industrial wastes, salt, and de-icing compounds from city streets, together with other materials upsetting the total dissolved solids, nutrient, and organic matter balance of aquatic ecosystems. Toxic chemicals like pesticides, spilled or dumped into a stream or lake often have a quick and devastating impact. When present in sublethal amounts, they may be accumulated by organisms and passed to the next trophic level in lethal amounts. Surface water contamination by oils or detergents can cause heavy mortality of water birds, because it severely damages their feathers.[10]

3. Impacts on water flow or current

Channelization and damming for flood control, erosion control, irrigation, or navigation are responsible for most current alterations. Channelization changes a stream from one having a variety of flows in pools, riffles, and backwaters to one having a straight and swift (laminar) flow. Reduced flow diversity causes a reduction in species diversity since fewer niches are available for habitation.

Converting free-flowing tributary streams to storm sewers essentially eliminates them as habitats for desired forms of aquatic life. Sudia[11] reported that a section of the Rock Creek watershed in Maryland, when a rural area in 1913, had 64 miles of natural flowing streams, but had only 27 miles of streams flowing through a heavily populated suburban Washington, DC area by 1966. Were it not for Rock Creek Park in the District, there would be even fewer miles of free flowing streams. This represents a substantial loss of fish habitat and a significant loss of opportunity for water-based activities.

Many areas in the Los Angeles basin which formerly contained populations of unarmored threespine stickleback *(Gasterosteus aculeatus williamsoni)*, an endangered fish, now contain none, and few if any other native fishes.

[9]W. T. Edmondson and C. C. Anderson, *Artificial Eutrophication of Lake Washington*, pp. 47-53 in Limnology and Oceanography, Vol. 1 (1), 1956.

[10]G. L. Choules, W. C. Russell, and D. A. Gauthier, *Duck Mortality from Detergent-Polluted Water*, pp. 410-414 in Journal of Wildlife Management, Vol. 42 (2), 1978.

[11]T. W. Sudia, *Man Nature City—the urban ecosystem*, USDI National Park Service, Urban Ecology Series, 1, 1974, 18 pp.

[redacted]

Urbanization, with stream channelization destroying the quiet backwaters and side streams required by sticklebacks, is one of the causes listed for the decline of this fish.[12]

Water flow also is affected by urban watershed management. Removing terrestrial vegetation by covering previously vegetated land with buildings and pavement speeds runoff and facilitates entry of pollutants into water bodies. Natural vegetation holds and slowly releases rainwaters, moderating water fluctuations during wet and dry periods.

In the Piedmont province of Maryland, stream quality was impaired when watershed impermeability owing to urbanization reached 12 percent, and became severe when imperviousness reached 30 percent. Five of nine urban streams sampled were completely devoid of fish life; in three of four urban streams where fish were collected, the dominant species was the blacknose dace, a fish that can tolerate a wide range of environmental conditions.[13]

Draining a wetland not only destroys valuable fish and wildlife habitat at the site, but also eliminates the former values of the wetland for holding water and contributing to stream flows in dry periods.

[12]S. Sasaki, J. N. Baskin, B. Beal, J. A. St. Amant, C. Swift, and M. A. Bell, *Recovery Plan for Unarmored Threespine Stickleback, Gasterosteus aculeatus williamsoni an Endangered Fish,* California Dept. of Fish and Game, 1977, 49 pp. and appendices.

[13]R. D. Klein, *Urbanization and Stream Quality Impairment,* Water Resources Bulletin, Vol. 15 (3), Water Resources Association, 1979, 16 pp.

Presumably because of drainage, channelization, and other urbanization impacts, 14 of 21 amphibian and reptilian species present on the edge of Indianapolis between 1949-1958 had been eliminated by 1963-1964.[14]

River damming changes flow regimes both up and downstream. Reservoirs have relatively quiet and often deep waters which may cover former riffles of a stream. Downstream flow may change radically as water is released for hydroelectric generation. Fluctuation, both in the reservoir and downstream, makes habitation of shallow littoral or shore zones difficult. Plants are repeatedly desiccated or unable to establish under regular fluctuation and flooding, and fish nesting in the shallows most often abandon their nests, leaving eggs to die when exposed to air.

4. Urbanization effects on water temperature

Water temperatures are often affected by industrial and power plants, removal of streamside vegetation, and in other ways. Operation of pumped storage hydroelectric projects involves pumping of water from a lower reservoir to a reservoir at higher elevation, from which water is released through turbines back into the lower lake. This may change the temperature stratification and turnover regime of a lake and the life cycles of resident organisms.

[14]S. A. Minton, Jr., *The Fate of Amphibians and Reptiles in a Suburban Area,* pp. 113-116 in Journal of Herpetology, Vol. 2 (3-4), 1968.

Heritage Conservation & Recreation Service

Figure 5. *The littered and polluted Rock Creek at the point where it flows under an overpass on its way to the Potomac River.*

[redacted]

Willis J. Ridenour, USDA, Soil Conservation Service

Figure 6. *Effects of channelization and improper stream bank protection.*

Downstream releases from a reservoir can be warm, if the discharge is from the surface, or cold, if from the bottom. Some thermal changes have been drastic enough to permit trout to survive in streams which previously harbored only warm water fish.

Heat is added to water systems from power and industrial plants using water to cool condensers or for other purposes. The warm water "plumes" which form in the receiving waters affect the behavior of aquatic organisms and cause some mortality. Organisms acclimated to warm water in winter are likely to suffer stress during periods of plant shutdown. Some animals, particularly during winter, are attracted to, and may benefit from, the warm water.

The giant Canada goose (*Branta canadensis maxima)*, once thought to be extinct, was rediscovered in 1962 on the cooling lake of a city power plant at Rochester, Minnesota. The lake was kept from freezing by warm-water effluent from the plant.[15]

The Florida or West Indian manatee (*Trichechus manatus*), an endangered species feeding on aquatic plants, formerly used some of the south Florida springs and estuarine areas in the winter. Because of development, many of these areas are no longer suitable for manatees, and many of those remaining show up in winter in the warm water discharge areas of Florida power plants. The manatee is susceptible to cold and many of them were found dead during the severe winter of 1976-77. Those in the vicinity of the power plants, however, were reported to have fared well.[16]

At numerous power plants in urban areas throughout the country, fish are attracted to warm water plumes, where they provide considerable opportunity for angling. At the Hunters Point Power Plant on the India Basin on San Francisco Bay, where public access was permitted to virtually all shoreline adjacent to the plant, angler numbers were heaviest in areas next to the thermal dis-

[15]H. C. Hanson, *The Giant Canada Goose*, Southern Illinois University Press, Carbondale and Edwardsville, Illinois, 1965, 226 pp.

[16]Frank Graham, Jr., *A New Hand in the Wildlife Business—big public utilities, those old archenemies of conservationists, are playing an increasingly important role as habitat managers*, pp. 94-101, 103-104, 110, 113, in Audubon, Vol. 81 (3), 1979.

U.S. Fish and Wildlife Service

Figure 7. *This piling provides attachment sites for various invertebrates.*

charges. Surfperches were taken primarily near the terminus of two pipe conduits providing a constant flow of intake cooling waters, while striped bass were associated with areas of greatest thermal discharge.[17]

Watershed clearing may also change stream thermal regimes. A forested watershed generally releases cooler water because of the evaporative cooling provided by terrestrial plants. A bare watershed is subject to more intense solar heating, which ultimately warms stream water, and can result in the elimination of some stream organisms and their replacement by others.

5. Effects on sunlight penetration into waters

Removal of vegetation overhanging a stream, in addition to increasing water temperature, also increases sunlight penetration because the shade vegetation provides is absent. Silt introduced as a consequence of construction and other soil disturbances on the watershed probably has greater significance. In turbid waters, sunlight penetration, and hence photosynthesis, is reduced, which results in less primary and less overall production. However, because the damming of turbid rivers causes suspended solids to settle in the reservoir's slow currents, river water below the dam is clearer than it would be otherwise. Species which have adapted to turbidity, for example the Colorado squawfish in the naturally turbid Colorado River, therefore, are at a competitive disadvantage. Species preferring clear water may invade sections of the river below a dam.

6. Effects on the substrate and physical configuration of water bodies

Impacts of stream channelization and dams have been briefly discussed. Silt is trapped in manmade impoundments and may dominate the substrate or bottom of these water bodies, particularly in small impoundments. A silt substrate, while sometimes highly productive of biomass, may bury rubble and boulders and reduce species diversity. Invertebrates such as midge larvae and carp that tolerate silty conditions often dominate the system as a result. Sediment deposition in reservoirs and settling basins can smother developing fish eggs and other bottom-dwelling organisms, can cause gill clogging, and also act as an abrasive. Additionally, some of the toxicants entering the water system are adsorbed onto soil particles. Control of accelerated erosion resulting in sedimentation is largely a matter of land use and application of erosion control measures on the watershed. Stream bank stabilization and quick revegetation of areas disturbed by construction activities in urban areas reduce erosion and sedimentation.

In five spring-fed streams near Atlanta, Georgia, dusky salamander populations were found to be inversely proportional to the degree of urbanization. Salamander populations were significantly affected by scouring caused by increased runoff and soil erosion in disturbed areas. Salamanders are important stream predators themselves and they, in turn, are a prime food of certain snakes.[18]

Thus, physical instability within stream habitats caused by urbanization can cause disruptions in the food web of stream communities. This emphasizes, again, the importance of maintaining streams in as natural a condition as possible during development.

7. Effects of manmade structures and artifacts

Dams constructed on rivers occupied by anadromous (migratory) fish like salmon and American shad block the passage of these fish, eliminating spawning areas upriver of the dams unless fish passageways (ladders) are installed or means for transporting the fish around the dams are implemented. Similarly, culverts may restrict the use of certain stream segments by other species, not just those that are anadromous. Bridges interfere somewhat with the free flow of water, but if properly designed and sited, they can reduce the need for stream channelization in urban areas. Their underwater supporting structures serve as cover for fish and as attachment sites for various invertebrates. In some urban settings, tires, shopping carts, and other discarded items may provide the only non-silt, non-sand substrate present. Mortality by automobiles in urban-suburban areas results in the loss of many water-dependent animals, especially when highways cross the breeding migration routes of such forms as toads, frogs, and salamanders. Strategically placed culverts or tunnels beneath the highway can reduce such mortality. Even city lights and the lights of highway vehicles have an effect on some aquatic species.

Underwater portions of tower and causeway structures associated with electric transmission lines crossing parts of San Francisco Bay are estimated to provide nearly a quarter-acre of living surface for basic food organisms of marsh ecosystems along every mile of transmission lines.[19] Above-water parts of the structures are used as resting or perching sites by cormorants, gulls, and other water birds. In the Charlotte Harbor area on the west coast of Florida, many osprey nests are found on artificial structures such as waterway channel markers, signs, utility poles, and other high structures rather than in natural situations.[20]

Increased development and overuse of coastal shorelines affect the large and threatened marine turtles which rarely come onland except to lay eggs. Development of coastal areas for industry and tourism has destroyed many of the turtles' nesting sites and the bright city and highway lights confuse turtle hatchlings, attracting them inland where they die.[21]

8. Invasion or displacement of aquatic habitat by urban development

Inland lakeshores and wetlands have been drained or filled extensively for housing and industry. In coastal and estuarine areas as well, bays and wet areas have been filled (Figure 8) and attempts have been made to drain extensive

[17] C. E. Steitz, *Angler Use and Catch Composition in the Vicinity of the Discharges of Hunters Point Power Plant*, Report 7712, 12-75, Pacific Gas and Electric Company, 1975, 42 pp. and 5 appendices, 156 pp.

[18] P. N. Orser and D. J. Shure, *Effects of Urbanization on the Salamander Desmognathus fuscus fuscus*, pp. 1148-1154 in Ecology, Vol. 53 (6), 1972.

[19] P. H. Arend, *The Ecological Impacts of Transmission Lines on the Wildlife of San Francisco Bay*, a report prepared by Wildlife Associates, Novato, CA 94947 for the Pacific Gas and Electric Company, 1970, 21 pp. + 10 plates and a map.

[20] R. W. Schreiber and E. A. Schreiber, *Observations of Ospreys Nesting on Artificial Structures in Charlotte Harbor, Florida*, pp. 5-7 in Florida Field Naturalist, Vol. 5 (1), 1977.

[21] U.S. Department of the Interior, *Sea Turtles to Be Added to Threatened List*, News Release, May 21, 1975, 2 pp.

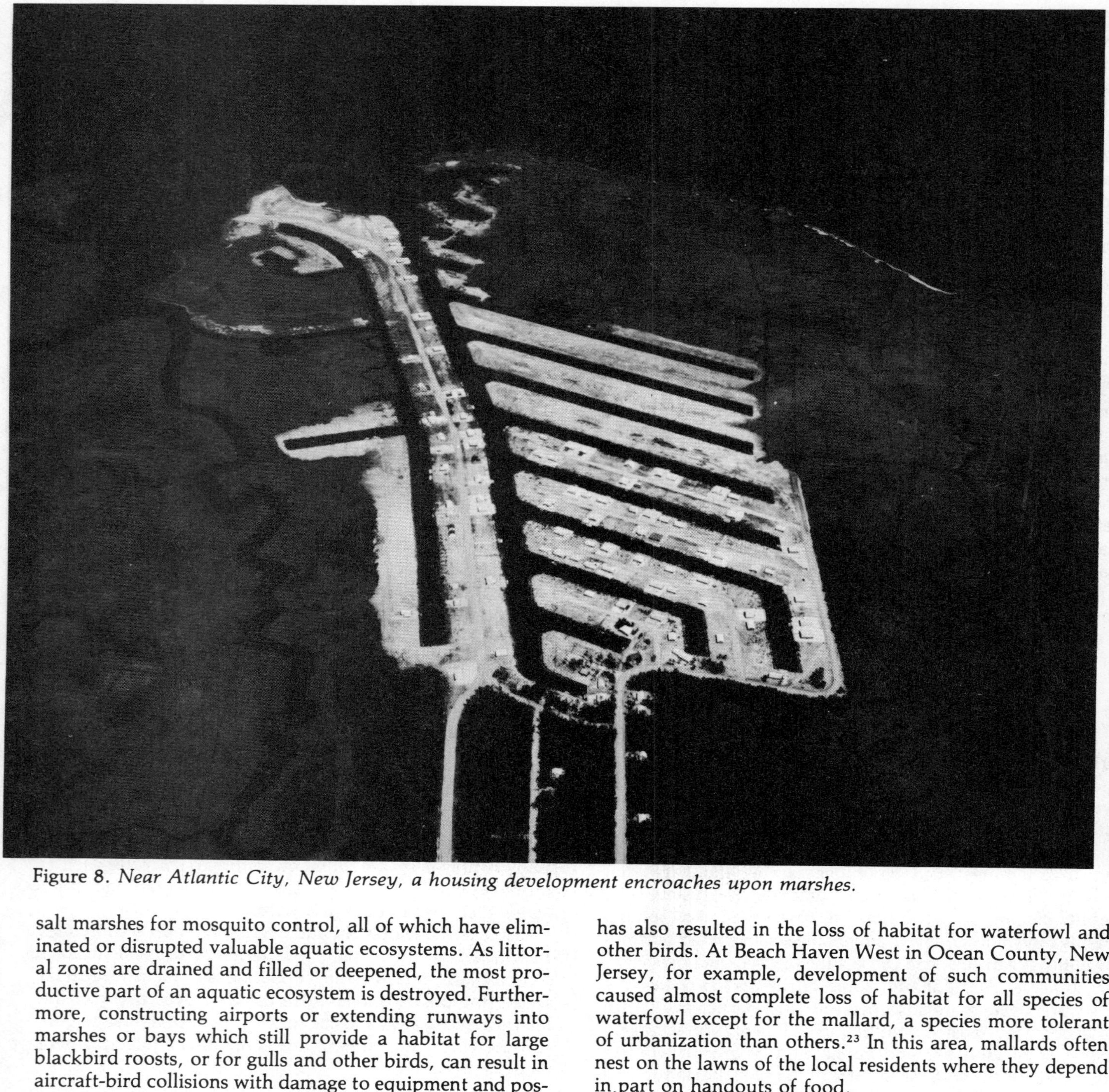

Charles D. Evans, U.S. Fish and Wildlife Service

Figure 8. *Near Atlantic City, New Jersey, a housing development encroaches upon marshes.*

salt marshes for mosquito control, all of which have eliminated or disrupted valuable aquatic ecosystems. As littoral zones are drained and filled or deepened, the most productive part of an aquatic ecosystem is destroyed. Furthermore, constructing airports or extending runways into marshes or bays which still provide a habitat for large blackbird roosts, or for gulls and other birds, can result in aircraft-bird collisions with damage to equipment and possible loss of human life.

In Florida an estimated 23 percent of good alligator habitat was lost in the development of six southern counties.[22] Because alligators are still present in the vicinity of housing developments, they may appear on urban lawns and golf courses as well as in urban marshes and impoundments. Occasionally they prey on pets; many people are afraid of them because an American alligator ranges in size to over 10 feet. Nuisance animals have to be removed at considerable expense. The conflict arises as a result of development in areas attractive to man and alligator alike.

Development of coastal lagoon housing communities has also resulted in the loss of habitat for waterfowl and other birds. At Beach Haven West in Ocean County, New Jersey, for example, development of such communities caused almost complete loss of habitat for all species of waterfowl except for the mallard, a species more tolerant of urbanization than others.[23] In this area, mallards often nest on the lawns of the local residents where they depend in part on handouts of food.

9. Creation of new aquatic habitat

In addition to the construction of large multipurpose reservoirs providing habitats for fish and other aquatic life, urbanization results in the creation of many other water bodies. Among these are recreation lakes and ponds, water supply reservoirs, sediment basins, sewage lagoons, wet gravel and borrow pits, and canals. Canals make it possible for mobile species like fish to invade other areas.

[22]L. D. Garrick and J. W. Lang, *The Alligator Revealed*, pp. 54-61 in Natural History, 86 (6), 1977.

[23]W. K. Figley and L. W. VanDruff, *The Ecology of Nesting and Brood Rearing by Suburban Mallards*, pp. 87-93 in Wildlife in an Urbanizing Environment—proceedings of a symposium, Noyes and Progulske, eds., University of Massachusetts, Amherst, MA, Holdsworth Natural Resources Center, Planning and Resources Development Series No. 28, 1974, 182 pp.

For example, construction of the St. Lawrence Seaway permitted invasion of the parasitic sea lamprey into the Great Lakes, contributing to the demise of an important trout fishery in those lakes. On the other hand, some canals provide good fishing and, in Florida, weed-filled canals are frequented by the West Indian manatee, which feeds on aquatic plants.

Many of the water bodies mentioned above constitute habitats for fish and other aquatic species and provide opportunities for fishing or bird watching. Sometimes harbor improvement or maintenance projects create habitats valuable for waterfowl or other aquatic species which contribute to recreational and aesthetic values in urban areas. Though a majority of the amphibians and reptiles native to Long Island, New York, have declined under urbanization pressures, new habitats suitable for such species as the eastern painted turtle and the common snapping turtle have been created by development of deep-water impoundments.[24]

The Outer Harbor Eastern Headland on the edge of the Port of Toronto, Canada was created for transportation purposes through use of material excavated for development in Toronto. Though not by design, the Headland has become a unique wildlife area in an urban setting. It is popular for recreation use, and many ring-billed gulls, Canada geese, common and Caspian terns, and other species nest there.[25]

The Port of London Authority dredges mud from the Thames and pumps it through a pipeline to huge embanked reservoirs or manmade lagoons which are attractive to large migrant flocks of wading birds. Though the lagoons rapidly become overgrown with vegetation after pumping operations cease, and hence unsuitable for waders, at one site the lagoons are operated on a rotation basis so that ideal conditions for these birds are almost permanently available.[26]

Implications for Planning and Management

As indicated in Chapter 1, an aquatic system has many factors influencing the biological community it supports.

[24]F. C. Schlauch, *City Snakes, Suburban Salamanders,* pp. 46-53 in Natural History, Vol. 85 (5), 1976.

[25]Arlene Gemmil, *Toronto's Outer Harbour Eastern Headland: The changing role of a transportation facility,* University of Toronto-York University Joint Program in Transportation Research Report 55, 1978, 86 pp. (University of Toronto, Room 219, 150 St. George Street, Toronto, Ontario, Canada M5S 1A1.)

[26](8) J. Harrison and P. Grant, *The Thames Transformed: London's River and its Waterfowl.* (Numbers in parentheses refer to previous footnotes.)

Heritage Conservation & Recreation Service

Figure 9. *A man-made pond in New York City's Central Park provides a habitat for fish and wildlife.*

Heritage Conservation & Recreation Service

Figure 10. *On the bank of Trout Creek, near Allentown, Pennsylvania, residents inspect the effect of chromic acid. Coming from an industrial facility upstream, this leak killed more than 1,000 fish.*

The more complex the habitat and the more niches provided, the more diverse the community. Generally, diverse communities are relatively stable because they contain a sufficient number of species to allow for substitutions in the various trophic levels as aquatic systems evolve naturally over time, thus protecting the flow and cycling of energy. Urbanization effects, when not immediately lethal, reduce the number of available niches, thereby reducing species diversity. Though total biomass may increase under changed conditions, as in eutrophic waters, species like bloom-causing algae and carp are not those most valued by man.

Urbanization effects on the aquatic environment are not all detrimental to fish and wildlife, however. We have noted that even without deliberate planning or management for fish and wildlife, many desirable species are present in urban and suburban areas, constituting a valuable part of the environment. The challenge, through planning and management, is to reduce or eliminate some of the negative environmental impacts of urbanization on fish and wildlife and to take advantage of opportunities enhancing these resources. A few examples illustrating how various communities have improved fish and aquatic wildlife conditions are cited below.

In the Detroit metropolitan area, productive urban fishing had been blocked by water pollution, limited access, and the focus of fishery management programs to more northern parts of the state. With a massive clean-up effort and enforcement of pollution control laws, water quality has improved and anglers are beginning to catch salmon and trout within the city limits. The clean-up effort got a big boost in 1968 when Michigan voters approved a $355 million Clean Water Bonding Program. On a dollar-matching basis, this program stimulated local, state, and federal agencies to produce more than $1 billion in clean water developments, with almost half the total utilized in the Detroit Metropolitan area. Additional hundreds of millions have been invested by industries in their own clean-up systems. This clean-up effort, with fish stocking and development of shore fishing facilities, has done much to boost sport fishing in the Detroit area.[27]

[27]N. E. Fogle, *Asphalt Angling,* Michigan Natural Resources, July-August, 1975, 4 pp.

The Mohawk River, flowing through the Utica-Rome, New York area, was so badly polluted that most sport fish species had been eliminated. Now that more than 75 percent of the discharge is controlled, highly sought-after sport fish like bass, walleye, perch, sunfish, and even trout have returned.[28]

Escambia, Pensacola, and East Bays in Florida, once supported speckled trout, oysters, shrimp, and even porpoises, but had been polluted so badly and were so excessively enriched by the early 1970s that hope of recovery was dim. To exacerbate matters, tidal circulation was impeded by a dense barrier of pilings from a railway bridge. Fish kills were rampant. Yet, with stringent pollution control efforts and the removal of unneeded bridge pilings, the system of bays was, by 1976, well on the way to a substantial recovery, and shrimp, oysters, and menhaden were all returning to the estuary.[29]

Within 35 miles of Philadelphia are 10 reservoirs built as part of the Neshaminy Creek Watershed Protection and Flood Control Project. This project, under P. L. 566, was sponsored by Bucks and Montgomery Counties, their boards of commissioners, and Pennsylvania conservation districts. One of the reservoirs, Lake Galena, is surrounded by a 1,500-acre county park whose temporary facilities for boating, fishing, and picnicking are already overused.[30]

At Pine Run Reservoir, recreational fishing by residents of the adjacent retirement community and nearby towns was enhanced by (1) constructing a series of fingerlike fishing peninsulas of dredged material at the reservoir's upper end, (2) deepening parts of it, and (3) stocking it with largemouth bass.[31]

At Columbia, Maryland, residents who had earlier regarded sediment basins as undesirable, later circulated petitions to save them because they enjoyed observing waterfowl, egrets, shorebirds, muskrats, and other wildlife using the matured ponds. In Maryland, responsibility for approving a developer's sediment control plan is assigned to soil conservation districts which obtain technical assistance from the Soil Conservation Service. The main reason for removing sediment basins is that some may be structurally unsafe because they were designed for temporary water storage rather than as permanent ponds. These shallow sediment ponds receive much greater use by wildlife than those that are deeper.

If such sediment basins are designed properly to assure dam integrity, or if they can be upgraded through improved piping systems and emergency spillways, their retention enhances water-dependent wildlife in urban areas, and the ponds continue to trap sediment. Mosquitoes, which may be a problem at early stages of development, are held in check by natural predators.[32]

The Suisun Marsh, about 85,000 acres of tidal marsh, managed wetlands, and waterways in southern Solano County, California, is the largest remaining wetland around San Francisco Bay. It is a fish and wildlife refuge of nationwide importance because it provides a wintering habitat for Pacific Flyway waterfowl, for striped bass—the most important game fish in the San Francisco Bay and Delta System—and for a variety of other fish and wildlife, including several rare and endangered species. Recognizing the threats to the area from potential developments and the need to protect this unique area, the California legislature passed the Nejedly-Bagley-Z'berg Suisun Marsh Preservation Act of 1974. This act directed the San Francisco Bay Conservation and Development Commission and the California Department of Fish and Game to prepare a Suisun Marsh Protection Plan "to preserve the integrity and assure continued wildlife use" of the marsh. The plan lists public hearings held for its preparation, assesses resources and values of the area, determines land uses and impacts, makes recommendations for carrying out the plan—including needed regulations, acquisitions, tax incentives, costs and funding—and includes maps.[33] This plan can serve as a model for other communities interested in protecting unique biological areas.

At Times Beach, Buffalo, New York, a 55-acre diked area on the Lake Erie shore of Buffalo's waterfront now has an abundance of wildlife, especially waterfowl and shorebirds. This area was created by the U.S. Army Corps of Engineers by filling a portion of the harbor with silt dredged from the Buffalo River. At least 186 bird species using this area have been identified, including many rare in the region. It seems likely that this unusual wildlife area can be preserved because of an agreement the Corps has with the city of Buffalo, and because of cooperation between Buffalo's Ornithological and Audubon Societies. These groups advise on dredged material disposition and vegetation management in the area. Bird watching and conservation education are suggested uses for the area.[34]

In the arid Southwest, wetlands are scarce. Much of the water available for recreation is in manmade reservoirs. Furthermore, institutional issues, the prior appropriation doctrine of western water law, and water allotment and delivery constraints complicate the possibilities for creating new wetlands for fish, wildlife, and recreation. To make helpful planning capabilities available for urban water and related land resource problems solution, the Albuquerque Greater Urban Area Resources Study (AGUA) was conducted by the U.S. Department of the Army, Corps of Engineers.[35] The study area, covering the main stem Rio Grande watershed from Cochiti Dam to the confluence of the Rio Puerco, contains about 30 percent of New Mexico's population. The river, which flows through Albuquerque, is an important recreation asset, and the river bottom contains the only major deciduous riparian woodland for 300 miles westward and more than 100 miles to the east. Because of the shifting sandy riverbed, together

[28]U.S. Environmental Protection Agency, *State of the Environment*, 6 (1), USEPA, Office of Public Awareness, (A-107), Washington, DC 20460, 1980, 33 pp.

[29](28) U.S. Environmental Protection Agency, *State of the Environment*.

[30]F. E. Bubb, *The World's Least Exclusive Yacht Club*, pp. 18-19 in Soil Conservation, 42 (8), 1977. (USDA, Soil Conservation Service, Washington, DC 20013.)

[31]G. N. Coller, *Fishing Fingers*, page 19 in Soil Conservation, 42 (8).

[32]C. W. Burdette and K. C. Gugulis, *Sediment Ponds Turn into Fresh-water Marshes*, 1978.

[33]San Francisco Bay Conservation and Development Commission (J. C. Houghteling, Chairman), *Suisun Marsh Protection Plan*, 1976, 48 pp. and maps. (San Francisco Bay Conservation and Development Commission, 30 Van Ness Avenue, San Francisco, CA 94102.)

[34]R. F. Andrele, *Times Beach: Buffalo's harbor for wildlife*, pp. 48-50 in Science on the March, Vol. 55 (4), 1976.

[35]U.S. Department of the Army, Corps of Engineers, *Albuquerque Greater Urban Area*, Information Bulletin No. 5, 1979, 7 pp. (U.S. Dept. of the Army, Corps of Engineers, P.O. Box 1580, Albuquerque, NM 87103.)

with variability and occasional flow cessation owing to water diversion, the only suitable aquatic habitat lies in the riverside drain and in a few small remaining marsh areas supplied by a constant ground water flow or river seepage.

Some study plan suggestions advised protection of existing wildlife areas from undue urban encroachment, preserving all existing wetlands (approximately 160 acres) within the study reach, and creating additional wetland acreage from borrow pits excavated for river levee rehabilitation.

As a result, the Corps of Engineers formed an Urban Study Group to address these questions.

Part II

Planning for Urban Fishing and Related Waterfront Recreation

Chapter 4

Planning for Urban Fishing

Status of Urban Fishing

By 1975, the U.S. Fish and Wildlife Service and 43 states had been involved with some form of urban fishing program, most of which had received local sponsorship.[36] These plans included urban water stocking, urban access programs, fishing piers, "kids only" fishing, fishing derbies, fishing day camps, swimming pool fisheries, senior citizens' fishing, fishermen's clinics, fisheries interpretive services, fee fishing, and more. About 74 percent of the states participated in stocking urban waters, 59 percent in park-pond fisheries, 41 percent in "kids only" fishing, 39 percent in urban area access programs, 33 percent in fishing derbies, and 26 percent in fishermen's clinics.[36]

Between 1932 and 1965, 1,545 community fishing lakes totalling 317,241 acres had been created for recreation in the North Central States. By 1960, this region spent substantially more money on lakes than any other region.[37] Even with the great increase in urban recreation lake acreage during the past decade, the extent of such waters is still quite small when compared with natural fresh water fishing areas, which total about 72 million acres in the contiguous 48 states, with an estimated 59.6 million acres in estuarine and coastal waters.[38] The problem, however, is that even in water-rich states, access to fishing areas for many urban center residents is limited. Most trips to significant recreational resources are made in personal automobiles or recreational vehicles; yet approximately 45 million Americans live in households without cars.[39]

Key problems with past programs for urban fishing centered on funding, planning, coordination, failure to use appropriate personnel, program publicity, and program evaluation. Fish managers need to manage sport fishing recreation rather than simply manage the fisheries.[36]

Though the U.S. Fish and Wildlife Service's direct involvement in urban fishing programs has declined since 1969-70, when it sponsored six pilot projects in cooperation with state conservation departments, municipalities, and private organizations, its continued interest and that of other federal agencies in urban fishing is reflected in the support given to preparation of this guide. The demand for urban fishing opportunities persists. Some of the programs initiated in 1969-1970 have continued while new programs have been initiated with the support of states, municipalities, and private organizations. Highlights of some of these pilot programs and other attempts to provide urban fishing are cited below.

Suggested Steps for Planners, Developers, and Decision-Makers

1. Review past and present urban fishing programs for quidance

The following examples provide insight about opportunities and constraints for developing urban programs.

(a) The cooperative pilot fishing program in St. Louis, Missouri, was initiated to provide outdoor recreational fishing experience for "inner city" dwellers. Five city park lakes previously stocked annually with carp, black bullheads, channel catfish, and green sunfish, were stocked in 1970 with carp and bullheads only. Carp and black bullheads, though often considered non-game fishes, were used for stocking because of their ready availability in the quantities needed. They could be caught on simple bait and proved to be suitable for the desired fishery type. Study results showed an estimated 140,487 fishing hours, that 77.1 percent of the fishermen ate the fish they caught, that about one-half walked to the lake, and that many of the fishermen were under 16 or over 60 years of age. Although there was no definite correlation between this urban fishing program and the incidence of crime in the fishing areas, many local residents thought the program was socially helpful.[40]

[36]M.W. Duttweiler, *Urban Sport Fishing: a review of literature and programs*, New York Cooperative Fishery Unit—Cornell University, U.S. Fish and Wildlife Service, and N.Y. State Dept. of Environmental Conservation cooperating, 1975, 52 pp.

[37]R. H. Stroud and R.C. Martin, *Fish Conservation Highlights 1963-67*, 1968, 147 pp. (Sport Fishing Institute, Suite 801, 601 13th St., N.W., Washington, DC 20005.)

[38]R. H. Stroud, "Recreational Fishing," pp. 53-66 in *Wildlife and America*, Council on Environmental Quality, Cosponsored by U.S. Fish and Wildlife Service, Forest Service, and National Oceanic and Atmospheric Administration, 1978, 532 pp. (For sale by Supt. of Documents, U.S. Government Printing Office, Washington, DC 20402, Stock number 041-011-00043-2.)

[39]U.S. Department of the Interior, Heritage Conservation and Recreation Service, and National Park Service, *National Urban Recreation Study: Executive Report*, 1978, 184 pp. (U.S. Government Printing Office, Washington, DC 20402.)

[40]A. Y. Ikeda, *A Study of the 1970 Urban Program in the City of St. Louis, Missouri*, M.Sc. Thesis (University of Missouri, Columbia, MO, 1971), 93 pp.

In this program, conservation education and communication with the youth of the city were important considerations. Groups of children were taught fishing fundamentals and then were permitted to fish in one of five intensively stocked city fishing lakes. The overall catch rate for the season was 0.45 fish per hour at a cost of $0.51 per fishing trip, or $0.61 per trip when costs were included.[41]

By 1979 the urban fishing program in St. Louis had nine park lakes regularly stocked with channel catfish, carp, and bullheads. In 1978 and 1979 alone, almost 77,000 pounds (35,000 kg) of fish were stocked in these lakes, with fishing pressure at urban lakes approaching 30,000 hours per hectare.[42]

(b) Kansas City began a similar urban fishing program in 1978. Administered by the Missouri Department of Conservation in cooperation with the Kansas City Parks and Recreation Department, it was financed by proceeds from Missouri's conservation sales tax. Fish stockings in several lakes and fishing clinics for urban youth are currently the main components of this urban fishing program.[43]

(c) In 1969, Lake Como, in Fort Worth, Texas, was stocked with adult size channel catfish and sunfish, a program costing $9,198.22, and which created an estimated 20,000 hours of recreation, 13 percent of which were attributed to children. A high percentage of the fishing was done by black citizens of the Como area.[44]

(d) In the 1960s, most (49) municipal water supply reservoirs in California, (83,000 surface acres), were being used by anglers. Calhoun[45] estimated that a typical 100-500 acre lake near Los Angeles put 50 pounds per acre of bass, sunfish, and catfish into anglers' creels annually. Fishermen usually stopped coming to a lake in large numbers when the average catch dropped much below a pound of warmwater fish a day. But their attendance increased dramatically when fish large enough to catch were stocked. Calhoun indicated that by stocking about 40 times the natural crop of trout in small streams near Los Angeles, a fortyfold increase in the amount of recreation over natural conditions could be generated. He envisioned that, eventually, there would be heavy sustained stocking in many municipal reservoirs, financed by daily fees, using trout in winter and other species in summer. At the same time, he felt that larger (over 500 acre) reservoirs could support major fisheries having natural fish crops. Often, however, engineers want few nutrients in the water, while biologists prefer a productive lake full of aquatic organisms to pro-

[41]D. J. Robinson, *Fishery Management Program—Urban Fishery Program, St. Louis, Missouri, Special Report*, 1970, 9 pp. + photographic section. (USDI Fish and Wildlife Service, Division of Fishery Services, Princeton, IN.)

[42]S. R. Alcorn, *Fishing Quality of the St. Louis Urban Fishing Program*, 41st Annual Meeting of the Association of Southeastern Biologists, March 26-29, 1980, Tampa, FL, 1980, In Press.

[43]P.J. Jeffries, *Evaluation of the 1979 Kansas City, Missouri, Urban Lake Sportfishery and Urban Fishing Program*, April 19, 1980, in litt.

[44]C.T. Menn, *Urban Fishing Program*. Job Progress Final Report, Federal Aid Project F-4-R-16, Job B-39, Texas Parks and Wildlife Dept., Bureau Sport Fisheries and Wildlife, Fort Worth Parks and Recreation Dept., and Office of Economic Opportunity, 1969, 17 pp.

[45]A. Calhoun, *Let's Have More Fishing Near Home*, pp. 13-14 in Outdoor California, 28 (4), 1967.

Heritage Conservation & Recreation Service

Figure 11. *This couple enjoys fishing and watching waterfowl on the Seattle waterfront.*

Figure 12. *These proud fishermen display carp caught while on a camping trip.*

duce bumper crops of fish. It *is* possible to have satisfactory drinking water and a good fish crop if biologists and engineers work together.

(e) In a later report, the California Department of Fish and Game[46] stated that the demand for fishing was so great near urban areas that the Department could not provide all the fish needed for a full-scale operation. Accordingly a new policy was proposed to set statewide standards for access fees and other aspects of cooperative fishing programs between the Department and local entities, encouraging matching programs in which local cooperators would purchase and stock fish to supplement the state's allotments. In this proposed policy, an urban reservoir was defined as one located within one hour's drive of a major metropolitan area. Programs, including the extent of state stocking, are described in the report for numerous metropolitan areas in California.

(f) Colorado's Division of Wildlife generally stocks salvaged warmwater fish and some catchable-size rainbow trout in its urban lakes. Salvaged warmwater fish, i.e., those which would die, be lost or are inaccessible to the general fishing public, are aquired from silting basins below irrigation reservoirs, reservoirs being drained, and certain private lakes and reservoirs where the Division has agreements with owners. Game fish commonly acquired include bluegill, green sunfish, black bullhead, yellow perch, and white and black crappies. Occasionally largemouth bass and channel catfish are obtained. Rainbow trout are stocked only in a few lakes and reservoirs along the Front Range. At study lakes located in Denver city parks—Berkeley, Garfield, and Houston—salvaged fish stocking, following carp eradication, could be accomplished at costs ranging from $0.40 to $0.60 per angler hour.[47]

(g) In the Phoenix and Tucson metropolitan areas of Arizona, 30 or more private and public reservoirs ranging in size from 0.4 to 34.4 surface hectares were reported by Wilbur.[48] He found private lakes to be real estate oriented, with fishing a low priority benefit. Public lakes were located in municipal parks, providing recreation ranging

[46]California Department of Fish and Game Inland Fisheries Branch, *Access Fees at Urban Reservoirs and Related Policy Problems*, Inland Fisheries Administrative Report No. 71-2, 1971, 22 pp.

[47]T. G. Powell, *Evaluation of Urban Lake Management Practices*, Colorado Div. of Wildlife, Fort Collins, CO, Project No. F-52-R, Work Plan III, Job 3, Job Final Report, 1976, 25 pp.

[48]R. L. Wilbur, *Urban Lakes*, Final report on Arizona Game and Fish Department's Project F-14-R-11 on Statewide Investigations, Work Plan 3, Job 01, 1976, 20 pp.

Photo: C. Franklin

Figure 13. *Recreational pier fishing is a popular pastime on Hatteras Island, North Carolina.*

Photo: T. Franklin

Figure 14. *Sport fishing is just a short walk up this ramp to the pier on Hatteras Island.*

from "come and feed the ducks" reflection pools and angling, to those allowing limited-size sail boating. Urban anglers were generally inexperienced, and while a majority of them were interested in catching anything they could, they expressed a preference for catfish and bluegills. Effects of fountains and waterfalls on oxygen and temperature levels were significant. They promoted a slight cooling of summer waters and, more importantly, they increased water circulation, helping to eliminate summer stratification of the water.

(h) Private entrepreneurs can provide fishing opportunities for urban and suburban residents. In a 1966 Illinois survey, 112 licensed operators of daily fee fishing ponds operating for one or more years, reported that 340,261 people had used their areas (246 ponds totaling 2,721.5 acres) for fishing, paying $512,269.50 in fishing fees. About 90 percent of these areas were located 50 miles or less from metropolitan centers.[49]

[49]A. C. Lopinot, *Illinois Daily Fee Fishing Ponds*, Illinois Dept. of Conservation, Division of Fisheries, Special Fisheries Report No. 14, 1966, 21 pp.

(i) Private industry also affords opportunity for urban fishing. For example, at its power plant located near Pittsburg, California, the Pacific Gas and Electric Company provides chemical toilets and trash receptacles for use by anglers fishing in the area of the plant's warm water discharge. In a 1975 survey, striped bass, white catfish, and carp accounted for 75, 15, and 6 percent of the total catch.[50] Fishing is permitted at many other urban plants in the United States.

(j) Private organizations like Trout Unlimited make important contributions to natural resource conservation and management in urban areas. For example, John T. Windell,[51] Chairman of the Colorado Council, Trout Unlimited, reported that by encouraging public hearings, presenting testimony, filing requests for action relating to minimum stream flow, and making habitat improvements,

[50]C. E. Steitz, *Angler Use and Catch Composition in the Vicinity of the Discharge of Pittsburg Power Plant*, Pacific Gas and Electric Company, Report 7712, 11-77, 1975, 123 pp.

[51]J. T. Windell, *Remaking a City Trout Stream*, pp. 22-25 in Trout, 19 (4), 1978.

Figure 15. *This streambank is protected by natural shrubs, principally elderberry, blackberry, amorpha, and herbaceous growth.*

Figure 16. *Clunei Park, located in the heart of Sacramento, California, is one of the many "spots" for complete relaxation for old and young alike in this capital city.*

Boulder Flycasters Chapter members helped make reaches of Boulder Creek, once avoided because of channelization and other urbanization impacts, popular again for trout fishing. This involved working with the U.S. Environmental Protection Agency, Colorado Division of Wildlife, Colorado Water Conservation Board, Boulder City Council and Administration, and the use of heavy equipment belonging to the Army Reserve. The work force included volunteers and heavy equipment operators from the Army Reserve. Financial assistance came from the City of Boulder, the Colorado Division of Wildlife, and the Mellon Foundation—Operation Restore.

(k) Chuck Woods[52] of Trout Unlimited's Maryland Chapter reported that the Anacostia Project Committee helped a Maryland National Capital Park and Planning Commission environmental planner make a Paint Branch watershed survey in eastern Montgomery County, an urbanizing area, with a view to protect native trout spawning areas. Such critical areas were identified and recommendations were made concerning Paint Branch Park boundaries revision, zoning and housing densities, road routing, and pollution abatement programs reflected in the master plan now regulating construction.

(l) In the seven-county metropolitan area of southeastern Michigan, which contains over half of the state's nine million people, trout fishing is minimal because only three or four small streams in this area can sustain trout year round. The nearest streams or rivers providing quality trout fishing are 150 to 200 miles north of the metropolitan area.

The Michigan Department of Conservation developed a program offering opportunities for trout fishing to residents of metropolitan Detroit by making use of the nearby Huron River which has flow and temperature conditions favorable for trout only during the spring months. As explained by Carl et al.,[53] the state stocks a river section with hatchery-reared trout of legal size or larger

[52]C. Woods, *Anacostia Project Report*, pp. 6-7 in Potomac Patuxent Conservationist, 4 (4), 1979.

[53]L.M. Carl, J.R. Ryckman, and W.C. Latta, *Management of Trout Fishing in a Metropolitan Area,* Michigan Dept. of Natural Resources, Fisheries Div., Fisheries Research Report 1836, 1978, 29 pp.

and imposes regulations for catching the fish from spring to fall. All fish caught during April and May must be released, but fishing regulations are liberalized as the season progresses. Between July 1 and September 30, fish can be caught and retained under the normal state-wide trout regulations. Sixty-four percent of the fishing took place in April and May during which each fish was caught approximately 2.35 times. Fishermen spent an average of $10.92 per trip, total benefits for the program were calculated to be $37,375, and total expenses were $3,708, a benefit to cost ratio of 10:1.1.

(m) The importance of sport fishing in estuarine and coastal areas near urban centers should not be overlooked. In a Long Island Regional Study led by the New England River Basins Commission with New York State, Connecticut, and Federal agencies cooperating,[54] it was estimated that the total economic return from sport fishing is probably more valuable than that from commercial fishing. The annual demand for salt-water sport fishing in the Sound, now estimated at 980,000 days, is expected to increase to 1,422,000 days by 2020. (See Recommended Readings.)

(n) In South Carolina, coastal pier fishing is popular in many areas such as Georgetown, Myrtle Beach, and Murrells Inlet, with many thousands of anglers visiting 11 or more piers weekly. At four piers surveyed, out-of-state people accounted for 57.2 percent of the anglers, in-state people 26.3 percent, and local residents 16.5 percent. Easy access to a fishing site and "the usual willingness of the spot (Leiostomus xanthurus) to bite a baited hook..." contribute to pier fishing popularity. Many piers remain open all night and a wide selection of baits and fishing equipment is available. Snack bars or restaurants are usually available at the piers also.[55]

Of 1,751 anglers interviewed, 72 percent caught at least one fish and 39 percent of the fish were either thrown back or given away. Approximately 75 cents of every dollar spent by the anglers went for lodging and food, 7 cents for gas and oil, and 18 cents for pier admission fees, bait, and tackle. The authors state: "Pier anglers injected $2.4 million directly into the local business economy. Of this amount, $1.3 million can be directly attributed to the presence of the pier industry."

2. Determine the attitudes and preferences of urban anglers

Planning and management require knowledge of urban anglers' demands, attitudes, and preferences, which can be obtained through studies, personal interviews, public meetings, or open hearings. Factors bearing on urban fishing's appeal and sites include: (a) ease of access (b) convenience (c) public safety (d) information availability (e) privacy (f) pollution level (g) fish species (h) fish quantity (i) fish size (j) facilities (k) natural beauty (l) water calmness (m) availability (n) alternative fishing site competition (o) alternative leisure activities' competition and (p) weather.[56 57 58] Costs and admission fees are also factors. Planners, managers, and pay-lake operators control many of these factors. A Massachusetts survey revealed that women anglers gave relaxation and nature enjoyment as their main reason for fishing.[59]

3. Review results of aquatic resources surveys and inventories

Planners and developers should have information on the extent, nature, distribution, location, and quality of existing water bodies in urban areas. Species, population, sizes, and fish characteristics occurring in these waters should also be known. Are fish game or nongame species? Are there enough fish for urban fishing? Are fish populations sustained by natural reproduction or dependent upon stocking? Is the habitat being polluted or sedimented? Answers and suggestions on ways to prevent harmful urbanization effects, rehabilitating degraded habitats, and managing fisheries can best be provided by biologists. See Part IV for technical and financial assistance sources.

4. Provide for preservation of existing high quality aquatic habitat

A development plan for preserving a high quality aquatic habitat in urbanizing areas should include provisions for avoiding construction on or near that habitat, for maintaining vegetation buffer strips along the waters, for quickly revegetating disturbed terrestrial sites, and for other action. See Chapter 3 and Part III for details.

5. Assess opportunities for providing additional habitat

Explore opportunities for developing new impoundments for fishing and other recreation both in urban and in urbanizing areas. When feasible, provisions for such impoundments in the plans will enhance recreational opportunities. Stroud[60] suggested that recreational fishery agencies identify feasible artificial lake sites in advance of development, and reserve them for future fishing use. He believes these sites would provide for lakes ranging from 50 to 1,000 surface acres (or more) having depths of not less than 10 feet and up to 50 feet maximum, and having drainage surface ratios ranging from at least 10 to not more than 25.

[54] USDI Fish and Wildlife Service and U.S. Department of Commerce, National Marine Fisheries Service, *People and the Sound: Fish and Wildlife*, 1975, 56 pp. and appendices. (New England River Basins Commission, 270 Orange Street, New Haven, CT 06511.)

[55] D. L. Hammond and D.M. Cupka, *An Economic and Biological Evaluation of the South Carolina Pier Fishery*, S.C. Marine Resources Center, Technical Report No. 20, 1977, iii + 14 pp. (S.C. Wildlife & Marine Resources Dept., P.O. Box 12559, Charleston, SC 29412.)

[56] M. W. Duttweiler, *Recommendations for evaluation of key elements of urban sport fishing programs*, Report 3 of 3 reports dealing with methodology for evaluating urban fishing programs, 1975, 21pp. (New York Cooperative Fishery Research Unit, Cornell University, Ithaca, NY.)

[57] G. H. Moeller and J. Engelken, *Fishermen Expectations and Pay-Lake Profits*, USDA Forest Service Research Paper NE-264, 1973, 5 pp. (N.E. Forest Experiment Station, Upper Darby, PA.)

[58] K. H. H. Beinssen, *Recreational and Commercial Estuarine Fishing in Victoria: A Preliminary Study*, Fisheries and Wildlife Paper, Victoria, No. 16, 1978, 40 pp. (Ministry for Conservation, Fisheries and Wildlife Div., 250 Victoria Parade, East Melbourne, Victoria, Australia 3002.

[59] I. C. Howard, *Opinions, Preferences, Satisfactions, and Importance of Women Anglers in Massachusetts*, pp. 32-34 in Fisheries, Vol. 4 (6), 1979. (American Fisheries Society, 5410 Grosvenor Lane, Bethesda, MD 20014.)

[60] (38) R.H. Stroud, *Recreational Fishing*.

Figure 17. *Providing access to fishable waters is a must. This photo was taken in Manhattan, New York.*

Reserving or acquiring reservoir sites will not be easy. Also, recreation lake values need to be weighed against those of stream habitats which they may replace. Certainly, however, some sites feasible for recreation lake construction exist. Information about recreation pond and lake maintenance, criteria for their design, construction, and legal and management considerations is available.[61] [62]

Fee fishing ponds or lakes constructed and operated by private entrepreneurs provide opportunities for urban fishing. Experience gained and reported on by operators of such lakes regarding requirements for successful operation has carry-over value for planning and managing public urban fishing programs. Based upon surveys of fee fishing operators in Illinois, Lopinot[63] established the following requirements:

(a) Location—within 25 miles of a metropolitan area.

(b) Access—easily accessible to the public and well marked by signs.

(c) Facilities—tackle and bait, seats and benches, comfort stations, picnic tables, camping areas, drinking water, shelters, club house/lodge, parking areas, food and beverages.

(d) Ponds—drainable, with good water supply source, 0.5 to 1.0 acre in size, and deep enough to prevent winter and summer kills (10 feet). Numerous ponds permit fish stocking variety. When heavy stocking is used, continuous running water through ponds helps prevent fish disease and die-offs.

(e) Fish supply—ample for "put and take" stocking.

(f) Fish species and stocking rates—for "put and take" fishing, with stocking rates recommended by Lopinot:

Species	Average Weight (lbs.)	Pounds per Acre to be Stocked Periodically
Carp	3.50	2,000-5,000
Channel Catfish	1.00	2,000-6,000
Trout	0.75	1,000-3,000
Bullheads	0.70	2,000-10,000

(g) Fishing fees—based on operational cost plus profit.

Similar facilities as suggested by Uhlig[64] and others included launching ramps (for larger lakes), fishing piers or platforms for the handicapped, a path along a dam's front berm, fish-cleaning huts or tables having water taps, garbage cans, night lighting, hard-surfaced paths, landscaped areas, and childrens' diversions.

6. Design facilities and incorporate provisions in the plan to facilitate multiple purpose use or reuse of selected urban waters

In addition to impoundments designed and managed expressly for fishing there are many other water bodies in urban areas providing opportunities for fishing. Some large multipurpose reservoirs constructed for flood control or electricity generation yield recreational benefits. Water supply reservoirs invite fishing and other water-based recreation. These larger impoundments, when equipped with water level control devices, can often be managed effectively for fish, wildlife, and recreation. Sediment ponds, wet gravel pits, borrow pits, wetlands or other impoundments created during urbanization, can provide opportunities for fishing, too. Often, their value can be enhanced by slight changes in size or shape during excavation. Sediment ponds may be designed for retention and periodic cleaning after construction. Fishery biologists can advise concerning water depths and fish species. (See Parts III and IV for additional information.)

[61] J. Tourbier and R. Westmacott, *Lakes and Ponds,* Technical Bulletin 72, 1976, 73 pp. (Urban Land Institute, 1200 18th St., N.W., Washington, DC 20036.

[62] U.S. Department of Agriculture Soil Conservation Service, *Ponds for Water Supply and Recreation,* Agriculture Handbook No. 387, 1971, 55 pp. (For sale by the Superintendent of Documents, U.S. Government Printing Office, Washington, DC 20402.)

[63] (49) A.C. Lopinot, *Illinois Daily Fee Fishing Ponds.*

[64] H. G. Uhlig, *Recreation Ready Reference,* USDA Soil Conservation Service, Northeast Technical Service Center, Broomall, PA, NETSC-Technical Note-Rec.-No. 1, 1977.

Recreational use of reservoirs reserved for domestic purposes has created some controversy and municipal reluctance, but studies show that these waters can be used for such activities when proper water treatment before domestic use is assured.[65] [66] Now many public water supply reservoir operators permit fishing.

There are opportunities, also, for using sewage treatment plant effluent as a water supply for recreation lakes, and research is under way to determine how the effluent may be used for fish production. Inadequately treated wastes from sewage plants, septic tank fields, or from storm sewers draining into recreation lakes can result in excessive enrichment and sometimes in increased bacterial concentrations in the waters.[67] [68] Anything that planners and developers can do to prevent situations where this might occur, e.g., residential developments having sewage treatment around recreation lakes, would be helpful in promoting fishable areas. On the other hand adequately treated effluent waters can supply the water needed for recreation lakes. As pointed out by Garrison and Miele,[69] in arid southern California, both economic and energy-conservation incentives exist for water re-use. They consider that the re-use mode resulting in greatest cost saving is groundwater recharge, but they recognize the need for investigating health-related aspects of water re-use and certain unknowns respecting re-use via groundwater recharge. They state, however, that presently constructed water-reclamation systems allow recreational water re-use and provide an excellent quality effluent for industrial, irrigation, and groundwater recharge re-uses.

Conceivably, therefore, water could be used for fishing before it is used for domestic purposes; it could be used for fishing after domestic waste water has been treated; and it could be used again for groundwater recharge.

7. Ensure that there is access to fishable waters

Providing convenient access to urban waters for anglers and other recreationists is an important consideration for recreation planning. Accessibility may depend on mass transit systems, or on personal vehicles for which parking areas are needed. New access roads or streets should be minimal, designed and routed for minimal sedimentation and pollution. Ways for minimizing urban roadway usage contributions to water pollution have been suggested by the Environmental Protection Agency.[70] Boat launching ramps, publicly owned river borders or lake shores having paths next to the water, fishing piers, docks, and bulkheads provide immediate access to fishing waters. In some situations, access requires land acquisition or leasing; in other cases, facilities at the end of streets dead-ending on a waterfront may suffice. Suggestions and examples on how various communities have provided access to fishing waters are cited below:

(a) To accommodate future fishing, Stroud[71] suggested constructing artificial fishing reefs and land piers, adding angler outwalks to bridges spanning coastal waters, capping and railing coastal jetties, and opening inaccessible or poorly accessible estuaries, coastal lagoons, bays, beaches, streams, rivers, lakes, and reservoirs to fishing.

(b) Grant County, in southwestern Wisconsin, bordered on the north by the Wisconsin River and on the west by the Mississippi, established a committee to review needs, seek new facilities, and explore ways of financing access developments. The county participated with the state of Wisconsin in a 50-50 cost sharing on access development and, as a result, public access sites were made available. Sites have adequate parking facilities, excellent boat-launching ramps, and are clearly marked with signs. Project maintenance has been good, with local conservation clubs assisting.

(c) Municipal and county officials in the metropolitan Detroit area are providing more means of access as part of their program for increasing urban-suburban fishing opportunities. For example, Fogle[72] reported that the people of Chesterfield township were building a large fishing pier into the north end of Lake St. Clair with support from the Michigan Department of Conservation. It will be 500 feet long, 10 feet wide, with lights, railings, benches, and a weather shelter. Under the pier and for 30 feet on each side, dredging provided a deep water area to attract fish. At Belle Isle, the fishing mecca in downtown Detroit, plans were well developed for constructing a very large fishing pier and bulkheads so anglers could reach water. Detroit's long-range goal is the development of a walkway corridor along its waterfront that is available to all.

8. Consider shelters for anglers in the planning process

Planners and builders should consider parking, picnicking, boat-launching, pier, dock, and facility shelters for anglers, since weather plays such an important role affecting a fishing expedition's success and enjoyment. Movable huts protecting ice-fishermen from winter cold should also be considered in urban fishing proposals.

In a 1949 creel census on 9,000 acre Lake Mendota, near Madison, Wisconsin, 6,600 ice fishermen were counted on six different dates. Ice fishing is a significant industry in the north and many fishermen use shelters as a protection against the cold.[73]

Oklahomans have pioneered in developing fishing docks which are heated in winter and air-conditioned in the summer. Sixty-eight enclosed fishing docks were reported at Grand Lake, at least 50 at Lake Texoma, several on most large U.S. Army Corps of Engineers reservoirs, and one on

[65] D. D. Bauman, *Perception and Public Policy in the Recreational Use of Domestic Water Supply Reservoirs,* pp. 543-554 in Water Resources Research, Vol. 5 (3), 1969.

[66] R. H. Stroud, *Recreational Use of Watersheds—Panel Discussion, Conservationists View,* pp. 1263-1270 in Journal of American Water Works Association, Vol. 58 (10), 1966.

[67] J. P. Baker and J. J. Magnuson, *Limnological Responses of Crystal Lake (Vilas County, Wisconsin) to Intensive Recreational Use 1924-1973,* pp. 47-61 in Transactions, Wisconsin Academy of Science, Arts and Letters, Vol. 64, 1976.

[68] D. G. Claudon, D. I. Thompson, E. H. Christenson, G. W. Lawton, and E. C. Dick, *Prolonged Salmonella Contamination of a Recreational Lake by Runoff Waters,* pp. 875-877 in Applied Microbiology, Vol. 21 (5), 1971.

[69] W. E. Garrison and R. P. Miele, *Current Trends in Water Reclamation Technology,* pp. 364-369 in American Water Works Association Journal, Vol. 69 (7), 1977.

[70] D. G. Shaheen, *Contributions of Urban Roadway Usage to Water Pollution,* Office of Research and Development, U.S. Environmental Protection Agency, EPA-600/2-75-004, 1975, pp. ix + 118 and 10 appendices comprising 228 pages.

[71] (38) R. H. Stroud, *Recreational Fishing.*

[72] (27) N. E. Fogle, *Asphalt Angling.*

[73] K. M. Mackenthun and Herman, *Creel Census,* Wisconsin Academy of Science, Arts and Letters, Vol. 39, 1949, Page 141.

many municipal water supply reservoirs in Oklahoma.[74] Lingenfelter and Summerfelt[75] estimated that, at Grand Lake, 18.3 percent of the catchable-sized crappie biomass—fish not usually harvested by fishermen—were taken by winter anglers fishing indoors on heated docks. They suggested that should warmwater fisheries accept additional fishing pressure with little danger to overall fishing success, then heated docks should be considered a means for increasing the sport fishing harvest.

9. Consider installation of fish shelters or other means of attracting fish to accessible waters

Urban fishing opportunities and catch rates can be increased by attracting or concentrating fish. As noted earlier, heated water plumes from electric power plants attract fish during winter and many electric utility companies permit public fishing on their properties. With negotiations, proper arrangements, and safeguards, more fishing opportunities at such plants probably could be made available. But there are other planned methods for attracting fish, some of which are mentioned below.

(a) Fish can be attracted by installing artificial structures on the bottom, on top, or suspended in midwater of a water body. The use and effectiveness of midwater artificial structures have been described by Hammond et al[76] and by Wickham et al.[77] Hammond and his co-authors pointed out that a midwater structure composed of six automobile tires lashed together and suspended from a steel cable, when used in conjunction with an existing reef composed of three steel-hull vessels off the coast of South Carolina, increased the concentration and availability of baitfish and pelagic (oceanic) gamefish. They observed that use of the artificial reef and the midwater structure together allows an angler to sample either or both groundfish and pelagic species without one angling technique interfering with the other. Artificial reefs or cover deliberately installed in coastal waters so they do not interfere with boating, attract fish and are often favorable fishing sites.

[74]R. C. Summerfelt, *Commercial Sport Fishing Enterprises and New Possibilities for Fishing Dock Operators,* Proceedings, 3rd Oklahoma Outdoor Recreation Conference, 1969, pp. 69-74. (Dept. of Agric. Econ. & OSU Extension Service, Oklahoma State University, Stillwater.)

[75]D.P. Lingenfelter and R.C. Summerfelt, *Angler Harvest in Heated Fishing Docks on an Oklahoma Reservoir,* Proceedings, Southeast Association Game and Fish Commissioners, Vol. 26, 1972, pp. 611-621.

[76]D. L. Hammond, D. O. Myatt, and D. M. Cupka, *Midwater Structures as a Potential Tool in the Management of the Fisheries Resources on South Carolina's Artificial Fishing Reefs,* S.C. Marine Resources Center, Technical Report, Ser. No. 15, 1977 iii + 19 pp.

[77]D. A. Wickham, J. W. Watson, Jr., and L. H. Ogren, *The Efficacy of Midwater Artificial Structures for Attracting Pelagic Sport Fish,* pp. 563-572 in Transactions, American Fisheries Society, 102 (3), 1973.

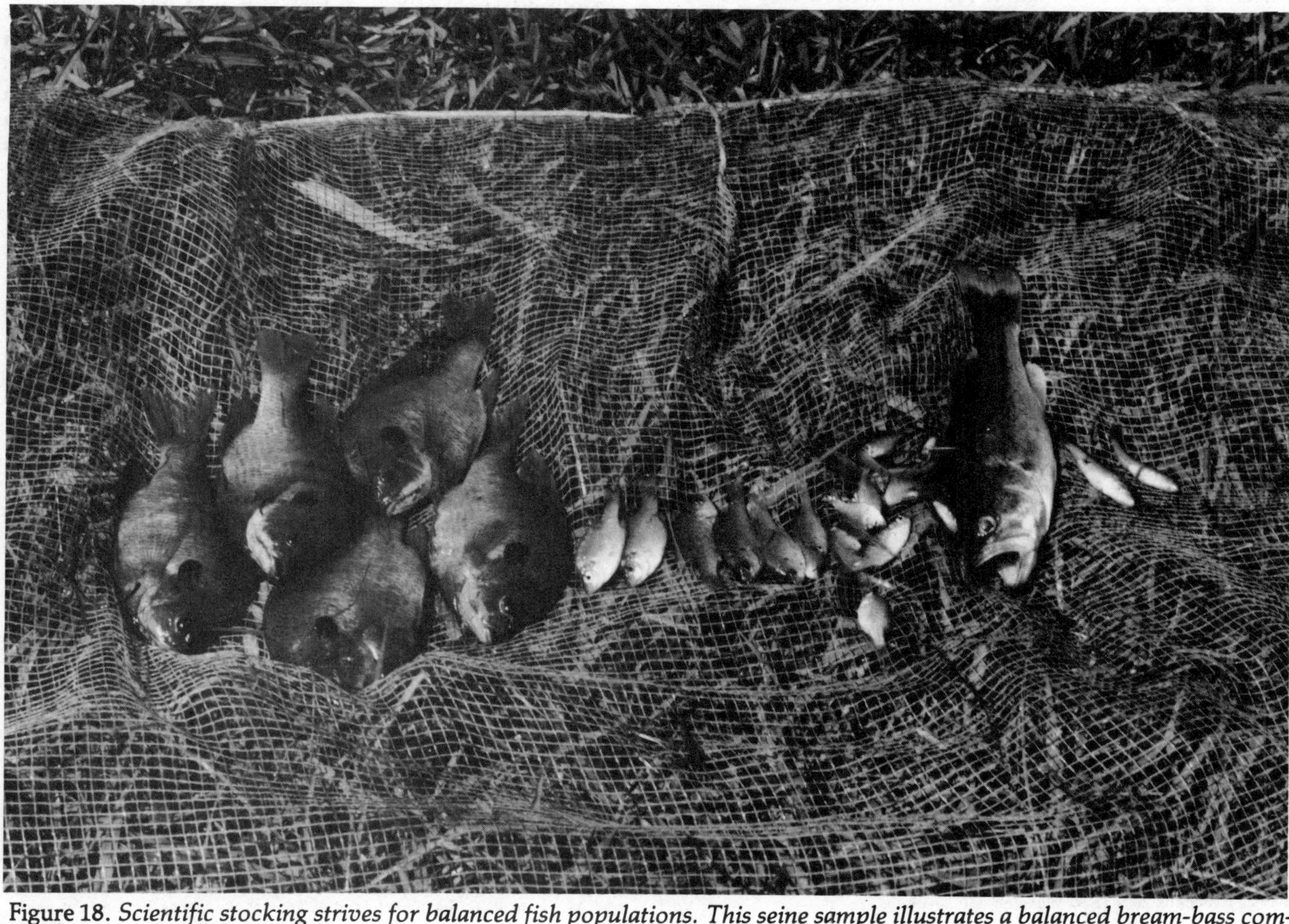

Photo: USDA, Soil Conservation Service

Figure 18. *Scientific stocking strives for balanced fish populations. This seine sample illustrates a balanced bream-bass combination in a warm water pond.*

(b) In fresh water lakes and reservoirs, brush, weighted or secured on the lake bottom provides attractive fish cover. On northern lakes which freeze in winter, brush structures placed at strategic locations on the ice surface will sink to the bottom when the ice melts, a convenient way of installing the structures.[78] Advice should be obtained from biologists about where and at what depth artificial underwater cover structures should be placed to be most effective.

(c) At intensely managed fishing docks, live minnows are sometimes suspended in the water in minnow traps as a lure to game fish.[79]

10. Review the plan

Review the plan in light of originally stated goals and objectives for enhancing urban fishing. Any newly obtained and synthesized data pertinent to solving specific problems or meeting specific goals should be included. Fishery biologists, economists, engineers, public and municipal agency representatives, and concerned citizens alike should be involved in the review. Questions regarding legal requirements, institutional arrangements, responsibilities of participating agencies and organizations, relation to other land use interests, economic feasibility and availability of funds, and biological feasibility should be addressed and changes made in the plan as necessary. (Refer to Part IV of the guide for information on possible technical and financial assistance sources.)

11. Implement the plan

Implementing plans for urban fishing requires well qualified personnel, coordination of effort, and the understanding and cooperation of the public.

12. Evaluate the results

Periodic evaluation of the plan should be made to determine its effectiveness and to provide for updating it as needed.

Fishery Management Considerations

Mention has already been made about fish stocking, fishing clinics, and other activities which relate to a fishery manager's work. These are discussed in more detail here because they are ways to enhance urban fishing and fishing enjoyment. However, planners and developers do have an important role in determining the nature and extent of urban waters requiring fish stocking.

1. Fish stocking

The introduction or stocking of fish in newly created urban impoundments or for "put-and-take" fishing is a necessary management measure. Before stocking a body of water, its suitability for fish and the need for stocking should be assured. Scientific stocking and regulation of fish populations can be very complicated; however, stocking successes and failures as discussed in the previous section and in the examples below may provide some guidance for future action.

(a) King[80] gives useful information to individuals who may have small cold-water ponds on their properties and who are interested in their management for trout fishing. Through the stocking and management measures he describes, he has had predictable fishing and excellent recreation on a one-fifth acre pond near Asheville, North Carolina.

(b) In connection with a pilot urban fishing program in St. Louis, a fish ratio of about 60 percent bullheads to 40 percent carp by weight appeared to be the highest obtainable during 1969-70. Carp should average 3 to 5 pounds each with a few large ones weighing up to 15 pounds, and the bullheads should weigh about 0.5 pounds each. Both are surplus fish subject to winter kill, and can cause serious fish and game management problems.

Stocking of inner city ponds at night may avoid interference from over-enthusiastic bystanders some of whom, in their excitement, want to take fish directly from the truck or haul them out of the water immediately after release.[81] However, fish stocking activities may provide opportunities for public education and recreation if properly planned.

(c) Largemouth bass and bluegills which may reproduce and maintain satisfactory populations in a pond lightly or moderately fished often cannot maintain themselves in a pond subjected to heavy fishing that is likely to occur near large metropolitan areas. This happened on Bluestem Lake, 25 miles from downtown Kansas City. In addition to heavy fishing, this 15-acre lake was usually turbid from wave action, making it difficult to manage for largemouth bass and bluegills. Accordingly, the fish population was removed and the lake restocked with 3- to 4-inch channel catfish fingerlings (3,000 per acre), 1,900 silvery minnows, and 700 fathead minnows—the minnows acting as supplemental forage to floating trout feed in pellet form. The experiment succeeded and fishermen were enthused and satisfied with catfish averaging 11.9 inches long weighing 10 ounces after one summer. Investigators reported that this lake supported 35,178 fishing trips (2,345 trips per acre) in 253 days of fishing. With a daily fee of 50 cents per person, this program netted $10,000 ($667 per acre).[82]

(d) Stocking lakes already populated by desirable species sometimes results in larger catches by fishermen, especially during the year of the stocking. Such stocking may be warranted in heavily fished areas, but usually it would not contribute much to sustained fish population production which is governed by the habitat's quality. Contributions stocked fish make to an angler's catch vary with the density of the native fish population. As an example, Lewis et al[83] reported on results for three stockings of marked,

[78]USDA Forest Service, *Wildlife Habitat Improvement Handbook*, USDA Forest Service Handbook 2609.11, Washington, DC, 1969.

[79](74) R. C. Summerfelt, *Commercial Sport Fishing Enterprises and New Possibilities for Fishing Dock Operators.*

[80]Willis King, *Homemade Trout Fishing*, Wildlife in North Carolina, Vol. 41 (10), pp. 13-15, 26, Vol. 41 (11), pp. 17-19, 1977.

[81](41) D. J. Robinson, *Fishery Management Program—Urban Fishery Program, St. Louis, Missouri.*

[82]M. L. Heman and F. M. Grogan, *The Development of an Intensively Managed Channel Catfish Sport Fishery in Missouri*, pp. 175-178 in Proceedings of North Central Warmwater Fish Culture-Management Workshop, R.J. Muncy and R.V. Bulkley, eds., Iowa Coop. Fishery Unit, Iowa State University, 247pp., 1971

[83]W. M. Lewis, R. C. Summerfelt, and A. Lopinot, *Results of Stocking Catchable-sized Warmwater Fishes in a Lake with an Established Fish Population*, pp. 235-238 in Transactions of the American Fisheries Society, Vol. 92 (3), 1963.

catchable-sized warmwater fish in a 30-acre lake (Izaak Walton Lake, Franklin County, Illinois) which already contained populations of bluegill, largemouth bass, and black crappie. Percent returns from spring stockings during the year in which the fish were stocked were crappies - 32 percent, bullheads - 28.3 percent, and bluegills - 26.1 percent. Further returns from these stockings during the following year averaged only 2.5 percent. From fall stockings the overall return the following year was 8.6 percent. In effect, therefore, these stockings represented a type of "put and take" fishing because the native fish catch the year after stocking compared to that before the stocking.

2. Placing more emphasis on panfishes and lightly fished species

Low catch levels for many panfish caught by recreational fishermen may have negligible effects on populations in comparison with environmental degradation, i.e., these fish can often withstand much more fishing pressure. Likewise, though carp are used in some urban fishing programs, lightly fished freshwater species like carp and suckers may have considerable recreational potential if interest in fishing for them could be generated.[84]

Municipalities and private organizations may wish to devote increased effort to educational and publicity programs to promulgate more use of such fish species.

3. Fishing clinics and similar projects

Fishing clinics and similar programs are important in the development of improved fishing ability and increased interest in conservation. Some state fish and game or conservation departments consider education in the art of fishing—and instilling in uninformed people some idea

[84](38) R. H. Stroud, *Recreational Fishing.*

Figure 19. *The National Park Service offers a fishing clinic at Hatteras National Seashore, North Carolina.*

Figure 20. *The Mississippi River as it flows past downtown, St. Paul, Minnesota.*

about the importance of aquatic resources—a necessary first step before engaging extensively in developing urban fishing programs.

(a) The Pennsylvania Fish Commission is sponsoring the Pennsylvania League of Angling Youth (PLAY). PLAY is an educational program helping youngsters become better anglers and safe boaters. It also hopes to make youngsters appreciate the aquatic environment and gives them a chance to show their concern for our rivers, streams, and lakes, and the life that inhabits them. Each subscriber, for $2 receives a membership card, a jacket patch, Fish Commission publications, a quarterly newsletter, and the use of a correspondence center answering any questions about fish, fishing, boating, and conservation.

(b) Howard[85] suggested that women's interest in fishing could be stimulated by arranging clinics where women and girls are taught basic fishing techniques, use of fishing equipment, safety precautions, and water-oriented ecology. These clinics could be organized and conducted by fish and wildlife agencies and groups like BASS, Trout Unlimited, sportsmen's clubs, or fishing tackle makers.

(c) Some states, like South Carolina, have published attractive and informative guides used by recreationists interested in fishing, clamming, crabbing, etc.[86] An illustrated book produced by Goldstein[87] provides much useful information for pier fishermen.

(d) The American Fishing Tackle Manufacturers Association (AFTMA) has developed a "National Youth Fishing Program" to encourage recreational fishing among the nation's youth. AFTMA's "Leader's Guide and Instructional Kit" tells how recreation admnistrators can organize group fishing activities and teach basic casting and angling. More than 2,500 kits have been distributed so far. For more information contact: AFTMA Center, 2625 Clearbrook Drive, Arlington Heights, IL 60005.

[85](59) I.C. Howard, *Opinions, Preferences, Satisfactions, and Importance of Women Anglers in Massachusetts.*

[86]C. J. Moore (ed.), *A Recreational Guide to Oystering, Clamming, Shrimping, and Crabbing in South Carolina,* Undated, 58 pp. (South Carolina Wildlife and Marine Resources Dept., P.O. Box 12559, Charleston, SC 29412.)

[87]R. J. Goldstein, *Pier Fishing in North Carolina,* John F. Blair, Publisher, 1406 Plaza Drive, Winston-Salem, NC 27103, 1978, 126 pp.

Chapter 5

Planning for Other Water-Based Recreation and Waterfront Values

Fishing in urban-suburban waters is only one water-related recreation activity. Others, like boating and waterfront picnicking, either combine or conflict with fishing. Planners must also consider a waterfront's historic, environmental, recreational, and aesthetic values. The term "waterfront" not only refers to a city's central core bordering on lakes, rivers, or bays, but also includes streams, ponds, reservoirs, and wetlands near cities and suburbs. Therefore, we are concerned with the extent and quality of urban aquatic resources, with adjacent lands, land uses, and buildings, and with the recreation opportunities these resources afford.

Status of Urban Waterfronts and Water-Related Recreation

Interest in waterfront revitalization has increased dramatically over the last five years—from Boston to Sarasota, Duluth to Davenport—in short, all across the nation.

Harney[88] has presented a thought-provoking account on urban waterfront evolution, with some complex questions and options to address in waterfront revitalization plans. Waterfronts, once the hub of power and trade for most American cities, more recently have been considered the seamy side of America. With the industrial revolution's advance, trains and trucks moved goods overland, with a corresponding decrease in waterfront trade. The city began to expand away from the water's edge. Residential areas relocated elsewhere, too, leaving shipping, commercial fishing and water-related industries alone on the waterfront. Relatively few people used the waterfront, divorcing it from the rest of the community. Waterfront areas in many cities deteriorated, leaving many vacant or underutilized structures. Subsequently, buildings, docks, and even water quality declined.

According to the Second National Water Assessment by the U.S. Water Resources Council,[89] less than one-fourth of the surface water area in the conterminous United States was accessible and usable for recreation because of pollution or other restrictions. Presumably the percentage of urban waters available for recreation was much less. Some urban waters were so badly polluted that they supported essentially no desirable aquatic life forms. Swimming and other water contact sports were prohibited.

This assessment showed 2 billion water-related outdoor recreation activity occasions (participation by a person 12 years of age or older without relation to duration) and a projected 34 percent growth to 2.7 billion occasions by the year 2000. Evaluation of a number of water-related recreation problem issues resulted in listing the following in descending order of priority: (1) preserving free-flowing stream values; (2) retaining flood plains, coastal beaches, and wetlands; (3) improving water quality; (4) optimizing recreation opportunities as reservoirs; (5) providing public access to water; and (6) maintaining instream flows.

"Reviving the Urban Waterfront" also identified certain themes and issues common to many waterfronts which needed to be considered in waterfront revitalization plans. These were outlined as follows:

(1) Cinderella syndrome—typified by the existence of many vacant, deteriorated, obsolete, or underutilized structures as well as by the secondary status which the waterfront takes to the downtown or outlying areas of a metropolitan area;

(2) aesthetic and cultural potential—for both manmade and natural waterfront features and water areas;

(3) environmental concerns—water and air quality, wetlands protection, shoreline maintenance, erosion control, storm and flood damage control;

(4) competition of uses—land, water-dependent and water, residential, recreational, parks, walks, commercial, tourist, industrial, power generation, waste disposal, marina, ports, shipping, and transportation (all modes);

(5) legal and institutional constraints—riparian rights, multi-level jurisdiction (state/local/regional/federal), lack of coordination, and self interest;

(6) economic considerations—tax issues, funding types and amounts available to plan, assemble land, redevelop or restore environmental quality, and market conditions.

[88] A. L. Harney (ed.), *Reviving the Urban Waterfront*, Partners for Livable Places, National Endowment for the Arts, and Office of Coastal Zone Management, a part of the National Oceanic and Atmospheric Administration (NOAA) of the U.S. Department of Commerce, Undated, 48 pp.

[89] U.S. Water Resources Council, *The Nation's Water Resources 1975-2000: Second National Water Assessment*, Vol. 1 Summary, 1978, 84 pp. (U.S. Water Resources Council, 2120 L Street, N.W., Washington, DC 20037.)

The report also stated that waterfront plans must embrace a safe environment, multi-use of land and building areas, deteriorated and abandoned area recycling or renewal, methods for overcoming physical barriers like freeways, major thoroughfares, barbed wire and chain-linked fences, visual access, creation of parks, open spaces or plazas, and pathways along shorelines.

Preserving or rehabilitating water areas suitable for water-based recreation can be accomplished with somewhat less difficulty outside of developed and often dilapidated urban center waterfronts. Fortunately, with community support and cooperation with federal and state agencies, progress is being made to enhance waterfront recreation opportunities.

Recently, HCRS helped form an interagency action group interested in protecting and revitalizing urban waterfronts. Called the Urban Waterfront Action Group (UWAG), it is composed of federal agencies such as Interior, Commerce, Transportation, Housing and Urban Development, Corps of Engineers, and others, together with nonfederal groups such as Partners for Livable Places, the National Trust for Historic Preservation, and the National League of Cities—all brought together by an official Memorandum of Agreement on November 1, 1979. Through this agreement, members acknowledge that rundown or underused waterfront areas can be restored for many uses, including housing, commerce, industry, and recreation.

UWAG can render coordination and cooperation for cities. The group set four initial goals:

- to distribute a directory listing contacts at participating agencies and departments;
- to serve as a clearinghouse through which interested communities and individuals might get advice on acquiring federal aid;
- to simplify bureaucratic machinery governing applications and deadlines, making it easier to get and use federal aids; and
- to expedite delivery of available federal assistance.

A directory of UWAG participants is given in Appendix L.

Enactment of various environmental laws has done much to reduce harmful environmental impacts of new construction, to reduce the amount of continuing water and air pollution, and to clean up existing polluted waters—one of the basic requirements for revitalizing waterfronts. The National Environmental Policy Act of 1969 administered by the Council on Environmental Quality (722 Jackson Place, N.W., Washington, DC 20006) and the Clean Water Act (33 U.S.C. 1251 *et. seq.*), particularly Sections 314 and 404, administered by the U.S. Environmental Protection Agency (401 M Street, S.W., Washington, DC 20460) are especially pertinent. The Environmental Protection Agency and many others can cite success stories describing how various projects have contributed to waterfront recreation. A few examples are given below, but the reader is encouraged to contact agencies and organizations listed in Part IV or the Appendices for additional information.

Examples of Actions Taken to Enhance Waterfront Recreation

Recreation and open space values are being incorporated into most major waterfront redevelopment projects. Two examples, Boston and Baltimore, have parks and marinas in conjunction with "festival markets" together with other commercial and residential uses.

(1) In Denver, Colorado, the Platte River Development project, nearing completion, includes parks, boating facilities, and a 10-mile paved walk-bikeway.[90]

(2) Portland, Maine's waterfront, deteriorated since World War I to the point where vacant floor space in buildings amounted to 16 percent, and is now being rejuvenated. Plan objectives include preservation and restoration of this historic architectural area to profitable uses, preservation of a finger pier at the water's edge, improvement in transportation and parking, preserving and upgrading the fishing industry in Portland Harbor, providing boating facilities, including marinas, public landings, and municipal pier facilities, completing an open space pedestrian belt along the waterfront, and expanding a greenbelt into the parks and landscaped streets.[91]

(3) The Huron-Clinton Metropolitan Authority or "Metroparks," a regional park agency dedicated to outdoor recreation for more than 4 million people in five southeastern Michigan counties, has developed park facilities in the Huron and Clinton river valleys encircling Detroit on the north, west, and south. In one 4,500-acre Metropark, annual attendance is about 2 million, or about the same as at Yellowstone National Park. Michigan's Department of Natural Resources, the State Water Resources Commission, river communities, and the Federal Government all participate in the clean-up of rivers, natural lakes, and artificial lakes constituting the waterways for the Metroparks. Canoeing has been encouraged by establishing canoe camping sites, assigning free canoe guides, and publicity. Intensive use of such urban rivers helps alleviate social pressures in crowded urban areas as well as those on more remote and sensitive river systems.[92]

(4) The Water, Power, and Resource Service's Pueblo Dam and 7,375-acre reservoir are located on the Arkansas River just six miles upstream from the center of Pueblo, Colorado. The relative absence of water-based recreation opportunities in Pueblo, a city of about 100,000 inhabitants, makes the reservoir and downstream river segment especially important in meeting the urban demand for such recreation. Facilities for swimming, fishing, bicycling, wading, picnicking, hiking, nature study, camping, and boating are presently under construction.

(5) The Kansas City Power & Light Company and the Kansas Gas and Electric Company have signed public use agreements for recreation uses on 5,500 acres of company property east of La Cygne, Kansas. The agreements call for a 600-acre park to be built and managed by Linn County, Kansas. 4,900 acres, including 2,400 acres of the station's

[90] C. T. Delaporte, *Clean Water Bonus*, EPA Journal Reprint (June), USEPA, Office of Public Awareness, (A-107), Washington, DC 20460, 1979, pp. 21-22.

[91] (88) A. L. Harney (ed.), *Reviving the Urban Waterfront.*

[92] R. L. Bryan, *Canoeing Use of Huron-Clinton Metropark*, pp. 121-124 in Proceedings: River Recreation Management and Research Symposium, North Central Forest Experiment Station, Forest Service, 1977. (USDA, 1992 Folwell Avenue, St. Paul, MN 55108.)

Photo: Heritage Conservation & Recreation Service

Figure 21. *Citizens enjoy waterfowl and boating on this lake in Boston, Massachusetts.*

cooling lake, is to be devoted to a wildlife management area under the Fish and Game Commission. The park is being financed by Heritage Conservation and Recreation Service funds matching the land and water value made available by the companies. Proposed recreation opportunities and park facilities include picnicking, camping, and hiking, in addition to marina facilities for boating and a heated dock for winter fishing.[93]

Suggested Approaches in Planning for Waterfront Recreation

Basic approaches for maintaining or increasing other water-related recreation opportunities are to preserve existing water bodies suitable for such activities, to rehabilitate those that have become degraded, to develop new water areas, and to provide access and facilities people need. Suggested steps incorporating these approaches follow.

1. Assemble information on existing urban-suburban water bodies and waterfront facilities

Data collected should include information on: water quality and quantity; present fisheries, water birds, and other aquatic or water-dependent animals having recreational value; and presence or lack of adequate access and waterfront facilities. Part III of this guide suggests types of aquatic resources data needed and approaches for obtaining the data, while Part IV presents technical assistance sources for collecting and interpreting information on water, soils, fish, and wildlife.

When planning new developments in urbanizing areas, identify on maps and document the presence of pristine streams, ponds, or wetlands, as well as unique biologic communities. Also, with the assistance of engineers and biologists, identify sites suitable for new impoundments or wetlands, weighing recreation values expected against those that would be lost.

2. Determine the attitudes and preferences of area citizens for recreation

Through studies, public hearings, and the like, determine the attitudes, preferences, needs, and demands for water-based recreation. Findings from selected studies may provide guidance.

(a) Based on visitor interviews at four western Nevada lakes—Tahoe, Pyramid, Lahontan, and Rye Patch—Myles[94] determined effects on water-based recreation of water quality and other factors, including the following.

- (i) A recreation area was commonly chosen over others because it was closer or more convenient. A 70-mile round trip apparently was considered near. Visitors appeared to prefer particular sites out of habit.
- (ii) Travel to and from the lake was a pleasant part of the recreational experience.
- (iii) Large open bodies of water in forested or desert surroundings seemed more scenic to most people than smaller ones like Rye Patch or Lahontan.

[93]Kansas City Power & Light Company and Kansas Gas & Electric Company, *La Cygne Generating Station*, Kansas City Power & Light Company, Undated, 13 pp.

[94]G. A. Myles, *Effect of Quality Factors on Water-Based Recreation in Western Nevada*, University of Nevada Agricultural Experiment Station Publication B-24, 1970, 62 pp.

(iv) 70-degree water temperatures induced people to say they liked the water because it was warm. Very few, however, mentioned Lake Tahoe water, averaging 65°F, as being warm.
(v) More restrooms, trash collection facilities and better maintenance were desired at many of the recreation areas studied.
(vi) Swimming areas could be improved by markers between boating and swimming areas.
(vii) Most visitors did not mind crowds; in fact, they seemed to like them.
(viii) Reasons given for liking a favorite outdoor recreation area were good facilities, clean, shade and greenery, scenic, good for swimming, skiing, or camping, and much to do.
(ix) Time appeared to limit outdoor recreational activities more than money.

(b) A Kentucky study of factors affecting the demand for outdoor recreation by urban residents revealed that as of March 1965, there were 691 private and 392 public outdoor recreation areas in the state. Pond fishing was the most popular activity at private recreational facilities but picnicking was most popular in public ones. Picnicking areas, in most cases, were complemented by a fishing lake, historic site, or some other type of attraction. Picnic areas generally were located close to urban population centers or heavily traveled tourist routes.[95]

(c) Minnesota canoeists and kayakers prefer lakes rather than rivers. As of 1977, Minnesota residents owned 64,118 paddle canoes and 1,577 kayaks. Eight of every 10 canoes and kayaks were used on lakes. About a third of canoe and kayak owners used their craft only on lakes and about a tenth used them on rivers only. The investigator suggested that management agencies could provide more water-based recreation opportunities for the high proportion of owners who reside in Minneapolis-St. Paul area, thus reducing pressures on more remote and sensitive waters. Close-in small urban streams such as Minnehaha Creek in the southern suburbs and south Minneapolis, and Rice Creek in the northern suburbs permit people to experience near-natural environments despite high human population density. Lack of access for motorized craft often limit motor craft use on urban rivers but nonmotorized craft use could be promoted if instream pollution were curbed.[96]

(d) Gunn[97] observed that social and economic gains are abundant whenever the rich resource assets of urban river corridors are redirected from waste containers and carriers to places of beauty, repose, and recreational use. He identified two types of river recreation development—ribbon and node. The ribbon type treats a waterway as a parkway by providing an aesthetically pleasing setting for distances along the watercourse; the node type provides a concentrated land-water interface at one location. The node type is illustrated by the San Antonio River Walk. The ribbon type is being implemented in Wichita, Kansas, along the Arkansas River. Water stabilization is needed in both types. As examples of urban-rural recreation waterways, Gunn cites the 425-mile long Trent-Severn-Rideau waterway in Ontario, Canada, and the 524-mile Barge Canal project in upstate New York. In the latter example, the canal passes through 21 counties, two-thirds of which are highly urbanized areas. Portions of this waterway are already overused, needing boating, camping, day-use, fishing, winter, and trail facilities.

(e) A Technical Note, "Recreation Ready Reference," published by the Northeast Technical Service Center of the Soil Conservation Service, contributes many rules of thumb helpful in recreation planning. It suggests, for example, that depending on shoreline configuration, water quality, depth, boat regulations, and policing, there should be about 15 acres per 10 or more horsepower boat for waterskiing on ponds, lakes, or reservoirs having average amenities. Boats having a motor of 10 horsepower or less, rowboats, canoes, and small sailboats, need 6 acres, while 2 acres suffice for a small fishing boat. Rating scales are provided, too, for shore fishing, camp sites, hiking, picnicking, and swimming. Other planning and design criteria are suggested for such facilities as launching ramps.[98]

3. Provide for preservation of unique aquatic-biologic communities

Provisions should be made in the plan for preserving pristine or unique aquatic systems identified when inventorying sites subject to development (see Step 1 above and Chapter 4). Threatened or endangered species and their critical habitats are protected under the Endangered Species Act administered by the U.S. Fish and Wildlife Service.

4. Do everything possible to minimize sedimentation and pollution

Planners and developers can help minimize sedimentation and pollution through careful siting of roads and structures, prompt reseeding of areas disturbed by construction, and retaining buffer strips of vegetation next to water bodies.

5. Encourage clean-up of degraded waters and waterfronts

The U.S. Environmental Protection Agency (see Appendix B for a list of regional offices) and state environmental protection agencies or their equivalent play a very important role in water pollution control. County and municipal officials, civic leaders, and others should be aware of the various programs that can be financed, at least in part, by EPA. Where available, EPA Basin Planning Documents provide a sound basis for further planning, and serve as helpful guides in identifying nonpoint source and other pollution problems requiring rectification within a drainage basin. Restoration of fishery potential is a sound basis for lake restoration under Section 314 of the Act. The EPA

[95] J. D. Wright, *Factors Affecting Demand for Outdoor Recreation by an Urban Area,* M.Sc. Thesis (University of Kentucky, Lexington, KY, 1966), 133 pp.

[96] Earl C. Leatherberry, *Minnesota Canoe and Kayak Owners: their characteristics and patterns of use,* USDA Forest Service, North Central Forest Experiment Station, St. Paul, MN, Research Paper NC-171, 1979, 8 pp.

[97] C.A. Gunn, *Urban Rivers as Recreation Resources,* pp. 19-26 in River Recreation Management and Research Symposium proceedings, USDA Forest Service, North Central Forest Experiment Station, St. Paul, MN, Technical Report NC-28, 1977, 455 pp.

[98] (64) H. G. Uhlig, *Recreation Ready Reference.*

receives annual reports from the states on the nature and extent of water quality in state waterways; these would be helpful to any planner in a particular region. Additionally, Section 404 of the Act relates to programs affecting water-based recreation, and the Section 201 facilities plan should be helpful from the standpoint of controlling sewage pollution. Cleaning up polluted urban waters often makes it possible to fish or swim where formerly boating only had been permitted.

Much useful information is contained in EPA publications, "Clean Lakes and Us"[99] and "Our Nation's Lakes."[100] Lake restoration is dealt with in the latter publication, in the proceedings of a national conference held in Minneapolis in 1978,[101] and in publications by Dunst et al.[102] and Nelson et al.[103] In addition to addressing ways for improving habitat in western reservoirs, the handbook by Nelson et al. deals with stream habitat improvement, a subject considered also by the USDA Forest Service[104] and by Barton and Winger.[105] Planners and developers should recognize that there are many ways to improve lake and stream habitats for fish and wildlife in addition to pollution control. It is suggested that they consult with biologists for guidance, e.g., in creating islands within lakes for added diversity.

[99] G. Gibson, L. Klessig, S. Nichols, and J. Peterson, *Clean Lakes and Us*, EPA 440/5-79-021, 1979, 37 pp. (U.S. Environmental Protection Agency, Washington, DC 20460.)

[100] U.S. Environmental Protection Agency, *Our Nation's Lakes*, EPA 440/5-8-009, 1980, 58 pp. (U.S. Environmental Protection Agency, 401 M Street, N.W., Washington, DC 20460.)

[101] U.S. Environmental Protection Agency, *Lake Restoration—Proceedings of a National Conference*, August 22-24, 1978, Minneapolis, Minnesota, EPA 440/5-79-001, 1979, 254 pp.

[102] R. C. Dunst, S. M. Born, P. D. Uttormark et al., *Survery of Lake Rehabilitation Techniques and Experiences*, University of Wisconsin and the Department of Natural Resources, Sponsored by the Upper Great Lakes Regional Commission, Dept. of Natural Resources, Madison, WI, Technical Bulletin No. 75, 1974, 179 pp.

[103] R. W. Nelson, G. C. Horak, and J. E. Olson, *Western Reservoir and Stream Habitat Improvements Handbook*, FWS/OBS-78/56, 1978, 250 pp. (USDI, Fish and Wildlife Service, Office of Biological Services, Drake Creekside Building, 2625 Redwing Road, Fort Collins, CO 80526.)

6. Provide for creation of new water areas where feasible

For those identified in the resources inventory as having potential for creating new aquatic resources (see Step 1), include appropriate designs for such developments. References cited in Chapter 4 for construction of recreation impoundments are relevant here.

7. Encourage use of areas not currently used for recreation

Multiple use of large reservoirs for recreation and other purposes is common, but there are other possibilities for providing recreation opportunities in urban areas. Through design changes, sediment basins created during construction can be retained and managed for fish and aquatic wildlife after construction, and, with slight changes in configuration, borrow pits and wet gravel pits can be rendered more productive.

[104] (78) USDA Forest Service, *Wildlife Habitat Improvement Handbook*.

[105] J. R. Barton and P. V. Winger, *Rehabilitation of a Channelized River in Utah: Hydraulic Engineering and the Environment*, pp. 1-10 in Proceedings of the Hydraulic Specialty Conference, Bozeman, Montana, Montana State University 1973.

Photo: Heritage Conservation & Recreation Service

Figure 22. *This stream is known as Muddy Creek because it carried so much mud before conservation practices were applied to the watershed.*

Figure 23. *Railings are needed on urban piers for public safety.*

Design opportunities at waste treatment facility sites as a means of enhancing urban recreation are discussed by Delaporte[106] and by Gerba and Hague[107] in connection with Section 201 of the Clean Water Act. Under this section the grant applicant must analyze potential recreation and open space opportunities in the planning of proposed treatment works. Though relatively few of the 13,000 wastewater treatment plants already constructed in this country provide for recreation, former HCRS Director Delaporte believes there is great potential at the 6,000 plants currently under construction if steps are taken early in the planning process to incorporate needed features. He believes that investing a little extra money—well below one percent of the total cost of a treatment plant—would enable these sites to become attractive and useful for public recreation and open space needs in addition to their sanitation function. Land and Water Conservation Fund grants matched locally on a 50-50 basis can be, and have been, used at wastewater treatment faciltiy sites for constructing access trails, bicycle paths and boat launching ramps.

An additional innovative use of an abandoned treatment plant is taking place in Bellingham, Washington. This project, funded by a UPARR grant, involves conversion of the facility to a salmon hatchery having extensive interpretive facilities. Located near the city's downtown area, it is truly an urban wildlife/recreation resource. Information is readily available from the HCRS Division of Urban Programs in Washington, DC. Information on Land and Water Conservation Fund grants is available from the State Outdoor Recreation Liaison Officer or from the Regional Offices of HCRS. (See Appendices D and E.)

[106](90) C. T. Delaporte, *Clean Water Bonus.*

[107]J. Gerba and B. Hague, *Recreation and Land Use: the Public Benefits of Clean Waters,* U.S. Environmental Protection Agency and USDI Heritage Conservation Recreation Service, Washington, DC, EPA 41/8, 1980, 43 pp.

The EPA publication by Gerba and Hague cited above contains much valuable information, not only on design opportunities for multiple use at waste treatment facility sites, but on other EPA programs impacting on urban recreation and the urban environment, including use of water clean-up by-products to improve land quality.

Other federal, state, and local agencies, private industry, and citizen groups also have important roles in providing recreation opportunities or facilities where none exist. Parks and even lakes are constructed on completed landfill areas. As examples, Johnson[108] has noted that Lincoln, Nebraska has two 50-year-old lakes built on an old city dump, which it uses exclusively for recreation purposes. Fishermen get one, boaters the other. The same community, with Lancaster County, has joined with a Soil Conservation District to build and manage some large flood control and erosion control lakes in Salt Creek Valley on the edge of the city. These are very popular places for recreation. Before designing and constructing lakes for fishing and aquatic contact sports on or near old city dumps, however, a thorough check should be made for any evidence of residual pollution.

8. Consider other needs and means for enhancing water-related recreation

Many people prefer non-water-related recreational activities near or in view of the water rather than boating, swimming, water skiing, or even fishing from the shore. These activities range from hiking, jogging, bicycling, and horseback riding to more leisurely nature walks, driving for pleasure, picnicking, or just sitting. People enjoy being near and looking at the water, the boats, the wildlife, or

[108]R. E. Johnson, *Examples of Urban Lake Management Problems,* in litt., dated April 7, 1980, to T. M. Franklin.

Photo: National Park Service, U.S. Department of the Interior

Figure 24. *A park ranger conducts a nature walk in Rock Creek Park, Washington, DC.*

whatever. From a planning standpoint, this means, again, that facilities for these activities should be incorporated in the plan and be designed and located in a way that fulfills their needs. Some suggestions follow:

(a) Paths (walking or bicycle) and roads built along, or in view of, urban waterways will be used by joggers and bicyclists. The tow path along the historic C&O Canal above Washington, DC is much used by joggers, hikers, and bicyclists. Roads having scenic views of water will be favorite routes for pleasure riding. Hikers, though content to use waterfront paths, will, like most nature walkers, prefer paths leading to fish and wildlife habitats and which offer seclusion. Some of these areas may include cliffs commanding outstanding views of a bay or ocean while others may cross ravines or water areas. Railings, signs, and other safety features are mandatory in such circumstances. Wetland areas afford excellent opportunities for viewing wildlife, but paths crossing these areas should be elevated to avoid trampling of vegetation and disruption of the habitat. Information on how to develop nature trails is available.[109 110] Well designed nature trails are an important conservation education tool, and consideration should be given to them in the planning of river greenbelts, waterfront parks, and other water-related open spaces. Generally, it is desirable to have separate trails for horseback riders.

(b) The interests of natural history buffs are quite varied. However, many recreationists having such interests concentrate on birds, which are relatively conspicuous. Opportunities for bird watching can be enhanced in several ways in addition to the development of nature trails. Motorists can view birds and other wildlife from the comfort of their cars, using turn-outs and parking areas along roads skirting or traversing a river, wetland area, or bay where large numbers of waterfowl or shorebirds concentrate. Turn-off examples include highways adjacent to the Horicon Marsh in Wisconsin with its large number of Canada geese, and a road along the state-owned refuge south of Anchorage, Alaska. In Anchorage, there is a system of linear parks along four of the creeks meandering through the city which are used for nature observation. On the Scioto River north of Columbus, Ohio, a variety of ducks can be seen.

(c) Establishment of wildlife refuges or sanctuaries in or near urban areas where habitat is preserved and wildlife is protected satisfies the recreational needs of wildlife observers. The Tinicum National Environmental Center near Philadelphia, the Blackwater National Wildlife Refuge near Cambridge, Maryland, the Chincoteague National Wildlife Refuge in Virginia, and the San Francisco Bay National Wildlife Complex are heavily patronized areas. At Blackwater, a road permits tourists to view many types of wildlife, with turn-outs and signs at strategic places to aid the visitor. A high observation tower is available, also, where visitors can get an expansive view of wetland areas and wintering geese. In some aquatic wildlife areas blinds facilitate observations.

[109]C. E. Mohr, *Environmental Study Areas: Wildlife Preserves,* Audobon Nature Bulletin, a part of set NB-9, 1961, 6 pp. (National Audobon Society, 950 Third Avenue, New York, NY 10022.)

[110](64) H. G. Uhlig, *Recreation Ready Reference.*

Figure 25. *A nature observation platform has been built in Adams Park, near downtown Atlanta, Georgia.*

(d) Viewing aquatic species like turtles can be accomplished in manmade pools near inner cities. For many years several turtle species in a small pool across from the U.S. Department of the Interior Building in Washington, DC, have given pleasure to hundreds of people. Kept in the pool during warmer months, the turtles are plainly visible when they crawl up on logs protruding from the water to sun themselves.

(e) More natural urban ponds with diversified vegetation along their banks create a habitat for many amphibians, reptiles and other forms of wildlife. Logs protruding from the water and rocks or brush piled along the banks furnish both cover and places for turtles, snakes, efts, newts, salamanders, dragonflies, and damselflies to bask. Tips on the management of amphibians and reptiles are available.[111]

(f) Clear water enhances recreation for many. It is needed, for example, for such activities as fish watching. Fish do not have to be eaten or even caught to furnish recreation. Many canoeists or boaters, proceeding leisurely along the shores and shallows of reservoirs, enjoy observing fish, fish nests, other aquatic organisms, and wildlife. Similarly, at the deep, clear water springs at Silver Springs, Florida, glass-bottomed boats allow people to view fish and other aquatic life clearly. Snorkeling and scuba diving are favorite pastimes for some people. Planning and management, obviously, must recognize how to maintain water quality and keep areas clear of undesirable vegetation and other obstruction interfering with such activities.

[111]T. R. Johnson, *Tips on the Management of Amphibians and Reptiles on Private Lands*, 1978, 15 pp. (Missouri Department of Conservation, P.O. Box 180, Jefferson City, MO 65102.)

9. Consider how to avoid conflicts among water-based recreation activities

Demand for water-based recreation and the relative scarcity of suitable close-in areas means keen competition among the varied activities. Conflicts arise, for example, when speed boaters and water-skiers, anglers, and swimmers, all attempt to use a pond or a small lake at the same time. Among several approaches taken to resolve conflicts are the following:

(a) Plan and develop separate areas for each major recreational use.

From the standpoint of planning, if space, physical conditions, and finances permit, one impoundment can be designed and developed for fishermen, with another for swimming or speed boating. Still another one can be designed and developed for use by waterfowl or other aquatic birds. Fishing ponds are not necessarily good for waterfowl and other water-dependent birds, as has been noted. Often, however, available resources do not permit single-purpose use so other means for solving conflicts are needed.

(b) Regulations

Regulations, if resources exist for their enforcement, can be effective in reducing conflicts. For example, on water supply reservoirs there is often concern that use of gasoline-fueled engines will result in water pollution. Hence, on many such reservoirs only electric motor boats or small motorless craft are permitted. However, a study on the effects of boating on lead concentrations in fish collected from lakes heavily used for boating in Kansas showed that lead concentrations in these fish did not differ significantly from fish collected in lakes where motor boating was prohibited. In this case it was concluded that lead levels apparently posed no public health hazard.[112] Noise of gasoline powered motors, waves created by fast moving boats, and other considerations such as wakes undermining shorelines and causing turbidity, may still warrant regulations prohibiting the use of such craft in some areas.

On lakes managed primarily as wildlife sanctuaries where people come to observe wildlife or commune with nature in a restful setting, motor boats can be outlawed because their noise disturbs not only the wildlife but the tranquility and enjoyment of people visiting the area. In shallow lakes having silty bottoms, it may be necessary to regulate against the use of motor boats to prevent stirring up the silt causing turbidity in the water. Where motor boats are permitted—and certainly their use is warranted and provides a great deal of recreation—regulations may be needed to keep the speed down in certain areas.

Boating speed limits and other control measures have been encouraged by various associations of boating law administrators, and by the work of such groups as the Sport Fishing Institute, the Outboard Boating Club, and the Izaak Walton League of America.[113]

From the standpoint of motor boat buffs, regulations may be required to prevent fishing boats from anchoring in the only stretch of water available to them for water-skiing. Likewise, certain types of artificial fish-attractant structures can be hazardous to motor boats. However, the type of structure—a subsurface unit in the form of a vinyl-covered cone—suggested by Wickham et al [114] is reported to prevent interference with other boating interests and to cause little damage if snared by a commercial fisherman's trawl.

(c) Zoning by time or space

Zoning facilitates the use of one body of water for several types of recreational activity. Jackson[115] suggests that activity zones on water can be delineated by grouping recreation pursuits into categories exhibiting similar density requirements and speed characteristics, thereby minimizing conflicts and competition. He proposes three activity zones, namely, shoreline, open water, and wildlife. The wildlife activity zone should be conterminous with shoreline land use zones, be liberal in dimension, and include both water and land. He states,

"The aims of the wildlife zone may be further reinforced by imposing, on sections of a shoreline, deed restrictions which limit beach alterations and improvements which riparian owners may perform. For example, lots could be sold with specifications of the shoreline and beach alterations that cannot be carried out. A deed restriction map, showing shoreline segments of a lake where aquatic plants must be left intact, where landfill and slope and material alterations are forbidden, etc., may be included as part of the Development Plan for each lake.

"...planning legislation, especially sections on the establishment of regional development plans, may in most cases provide the smoothest legal basis for the establishment of controls of water-oriented recreation. Since statutes differ considerably in their latitude of power, a pertinent procedure would be to amend the most appropriate act with a section which deals specifically with water-use zoning."

Zoning by time and place might be applicable in restricting or prohibiting use of an area in the immediate vicinity of a bald eagle aerie until the young have left their nest. Leatherberry[116] suggests that water surfaces should be zoned by time rather than space in resolving conflicts on lakes used by water skiers and boat fishermen. Thus, it would seem that the best approach depends upon needs and circumstances at a given location.

10. Include provisions for safety in designing and operating waterfront facilities

There is much that planners, developers, and managers can do to reduce the likelihood of drownings and other accidents. For example, sediment ponds in residential areas can be designed with gently sloping banks rather than steep banks down which playing children may slip and fall; quarries having high perpendicular walls, if used for recreation, can be fenced; and at scenic overview sites, fences or railings can be installed, with warning signs posted.

Apparently, liability is likely to be imposed only if the conduct of an operation, e.g., a fee fishing pond, is unreasonable and an accident or death occurs because the pond has no fence around it to keep young children (non patrons) from walking in and drowning. Other situations in which liability might be charged in case of an accident include: permitting children to play on fishing piers, where no rescue appliances or life guards are provided and falling from unprotected retaining walls, sharp banks, or as a result of depressions or obstructions in footpaths. The importance of incorporating safety provisions into planned recreation facilities and programs should not be underestimated.

11. Provide adequate access

Frequent mention has been made in this guide—mostly in connection with inland lakes and rivers—about the need for providing access to increase water-based recreational opportunities for urban and suburban residents. Obviously this need extends to coastal and estuarine areas, too.

In the Foreword to "Reviving the Urban Waterfront,"[117] Robert Knecht, Assistant Administrator, Office of Coastal Zone Management, stated:

"In 1976, Amendments to the Coastal Zone Management Act required state management programs to develop 'a planning process for the protection of and access to public beaches and other public coastal areas of environmental recreation, historical, aesthetic, ecological, or cultural value.' Although the language does not specifically refer to urban waterfronts, these are the very characteristics found in most in-town waterfront areas.

[112]David Oates, *The Effects of Boating Upon Lead Concentrations in Fish*, pp. 149-154 in Transactions, Kansas Academy of Science, Vol. 79 (3/4), 1976.

[113]R. Jackson, *Zoning to Regulate On-Water Recreation*, pp. 382-388 in Land Economics, Nov. 1971, University of Wisconsin Press, Madison, WI 53706, 1971.

[114](77) D. A. Wickham, J. W. Watson, Jr., and L. H. Ogren, *The Efficacy of Midwater Artificial Structures for Attracting Pelagic Sport Fish.*

[115](113) R. Jackson, *Zoning to Regulate On-Water Recreation.*

[116](96) Earl C. Leatherberry, *Minnesota Canoe and Kayak Owners: their characteristics and patterns of use.*

[117](88) A. L. Harney (ed.), *Reviving the Urban Waterfront.*

The regulations go on to say that 'special attention should be given to recreational needs of urban residents for increased shorefront access' echoing what is now an overall federal concern. Every federal agency, as well as state and local organization, must look hard at the potential inherent in their neglected urban waterfronts."

An example of approaches being used by federal agencies to facilitate access for recreationists is the Memorandum of Understanding (MOU) signed August 12, 1980 by the Heritage Conservation Recreation Service and the Federal Highway Administration. The MOU provides guidelines for the use of federal-aid highway funds available to build access ramps to public launching areas. The Federal Aid Highway Act of 1976 authorized access ramps to public boat launching areas, and required the two agencies to develop implementing guidelines.

Upon application by a state highway agency, federal highway funds may be used to build access ramps to public boat launching areas adjacent to bridges under construction or undergoing reconstruction, replacement, repair, or alteration on the federal-aid primary, secondary, and urban highways. Property next to the highway right-of-way where the boat launching area is to be located must be publicly owned at the time federal funds for the access ramps are obligated, and can be operated by an appropriate public agency.

If highway right-of-way allows joint development, a public boat launching area may be located within the right-of-way, so long as it does not interfere with the safety and utility of the highway.

A key element of the MOU is early notification to State Liaison Officers (SLO) of proposed bridge work and the opportunity for state funding. The SLO prepares a state's comprehensive outdoor recreation plan (SCORP).

As we have seen, however, much of the responsibility for providing access lies with municipalities, private industry, and organizations. At City Care,[118] a national conference on the urban environment held in Detroit, Michigan, workshop members concluded that public access should be a top priority. They stated:

"Various techniques for reducing land acquisition costs include the following: use of street ends, exploring small-scale designs; use of incentive zoning; tax incentives; special district designations; seeking donations, and requiring public access easement.

"Use of the above techniques has resulted in a greenbelt along the waterfronts of some cities, with public access guaranteed. In cities such as Detroit, private business involvement has been a significant factor in the creation of

[118]National Urban League, Inc., Sierra Club, and Urban Environmental Conference and Foundation. *City Care*, 1979.

Photo: T. Franklin

Figure 26. *The Herbert C. Bonner Bridge at Oregon Inlet, Hatteras Island, North Carolina, was designed with a catwalk to provide access and safety for fishermen.*

such greenbelts, since public easements have been granted along private property bordering the waterfront."

12. Consider acquisition of additional land and water areas

In addition to the ways suggested for acquiring land or reducing land costs in item 11 above, other methods are discussed by Whyte,[119] Strong,[120] Schmertz,[121] and Dunham.[122] Public acquisition is the surest way of maintaining control over land and water areas. The State of Missouri, through its successful Design for Conservation Program in which funds from state sales tax are designated for conservation purposes, is actively acquiring lands in conjunction with multiple use public works projects. Possibilities for coordinated acquisition of lands and waters and the roles of local, state, and federal agencies in such acquisition measures, including the distribution of Land and Water Conservation Funds administered by HCRS, are discussed by Gerba and Hague.[123] Various other sources of financial assistance for acquiring and developing land and water areas are described in Part IV.

13. Consider institutional and legal constraints

When planning and managing waterfront recreation facilities and programs, institutional aspects and legal requirements must be given careful consideration. From the standpoint of recreational rehabilitation of a river on a regional basis, redevelopment is complicated by differences between cities, i.e., physical setting, historic background, financial capability, policies, and objectives.[124] Much of what can be done with the river depends on land uses of the watershed. On the other hand, possible effects of water-based recreation developments on lands and waters owned by others must be considered. In some areas, especially tidal areas, land ownership and boundary disputes must be clarified before planning and development can proceed. In projects involving cooperation of several agencies, institutions, or organizations, it is important to understand each agency's responsibilities clearly. Experiences gained in the pilot fishing program conducted in a Fort Worth, Texas city park led one investigator[125] to recommend:

(a) any program involving more than one government, state, or city agency should be preceded by several intercooperator meetings, with all involved personnel in attendance and

(b) all agencies involved in such a program should have at least a general understanding of every phase of the project as well as a knowledge of the problems each agency will have to deal with in the performance of their commitments.

Environmental laws and ordinances and their enforcement at federal, state, and local levels regulate what can and cannot be done in aquatic resource development projects. They not only serve to protect the environment and require, for example, that certain standards of water quality be met, but many of them include authorization for funding assistance needed in planning and development projects. The environmental impact or assessment statements required by the National Environmental Policy Act of 1969 constitute a valuable natural resource planning tool.[126] Two Executive Orders issued by President Carter in 1977—E.O. 11988 establishing the federal position on floodplain management, and E.O. 11990 concerning the protection of wetlands—related to the preservation of areas valuable not only to fish and wildlife, but also for water storage, flood prevention, pollution abatement, and sediment reduction. Protection of such areas is needed before urban planners and others can optimize their use for fish, wildlife, and recreation. Some of the federal laws containing authorization for financial assistance to states and municipalities are discussed in Part IV.

Many states have enacted similar legislation and many municipalities have ordinances on the use or development of floodplains, sand and gravel deposits, wetlands, aquifer recharge, and other areas. Undoubtedly most planners are familiar with such laws and ordinances because of development constraints. A logical outgrowth of the planning and decision-making process which incorporates findings from environmental assessments and considerations for recreational resources, is the development of new ordinances or laws needed to implement the plan. A good example of this occurred in Collier County, Florida, which used the results of a comprehensive cooperative planning study to initiate a strong program regulating development and protecting its water system and marine resources through establishment of county ordinances.[127]

[119]W. H. Whyte, *The Last Landscape*, New York: Doubleday and Co., 1968, 376 pp.

[120]A.L. Strong, *Open Space for Urban America*, Prepared for the U.S. Department of Housing and Urban Development, 1965, 154 pp. (U.S. Government Printing Office, Washington, DC 20402.)

[121]M. F. Schmertz (ed.), *Open Space for People: Acquisition, Conservation, Creation, and Design*, 1975, 111 pp. (American Institute of Architects, 1735 New York Avenue, N.W., Washington, DC 20006.)

[122]A. Dunham, *Preservation of Open Space Areas: A Study of the Nongovernmental Role*, 1966, 101 pp. (Welfare Council of Metropolitan Chicago, 64 E. Jackson Blvd., Chicago, IL 60604.)

[123](107) J. Gerba and B. Hague, *Recreation and Land Use: the Public Benefits of Clean Waters*.

[124](97) C. A. Gunn, *Urban Rivers as Recreation Resources*.

[125](44) C.T. Menn, *Urban Fishing Program*.

[126]T. Dolan, IV and R. M. Maestro, *The Environmental Assessment Statement as a Natural Resource Planning Tool*, pp. 347-358 in Transactions, North American Wildlife and Natural Resources Conference, Vol. 40, 1975. (Wildlife Management Institute, 1000 Vermont Avenue, N.W., Washington, DC 20005.)

[127]J. Clark, *Rookery Bay: Ecological Constraints on Coastal Development*, 1974, 91 pp. (The Conservation Foundation, 1717 Massachusetts Avenue, N.W., Washington, DC.)

Part III

Integrating Aquatic Resource Considerations Into the Planning Process

Chapter 6

Site Planning

Integrating aquatic resource considerations into the planning process offers opportunities to (1) maintain or protect high quality or unique aquatic ecosystems, (2) enhance or rehabilitate disturbed systems, and (3) create new aquatic resources. This chapter suggests a procedure for incorporating helpful steps to ensure that water bodies in urban areas suitable for urban fishing or other desired uses are part of the plan.

Aquatic ecosystems are complex, involving not only the water and living organisms within or dependent on it, but also physicochemical components and their interrelationships with plants, animals, and land use on watersheds. Planning should focus on the entire aquatic ecosystem. Aquatic resource optimization depends on site planning that is sensitive and responds to the natural functioning of the aquatic environment. Figure 27 shows suggested steps for aquatic resources planning.

Step 1. Identify existing aquatic resources

An inventory of existing aquatic resources should be made early in the planning process and serves as the basis for planning and management efforts designed to protect or restore an aquatic resource. It should focus on the following:

(a) Delineation of aquatic resources and description of their structural and functional characteristics

Specificity and topics covered in an inventory can vary considerably, but the more detailed it is, the more useful it is. An inventory simply identifying the name and location of a lake or wetland is less useful than one giving size, physico-chemical-biological properties, and relationship to

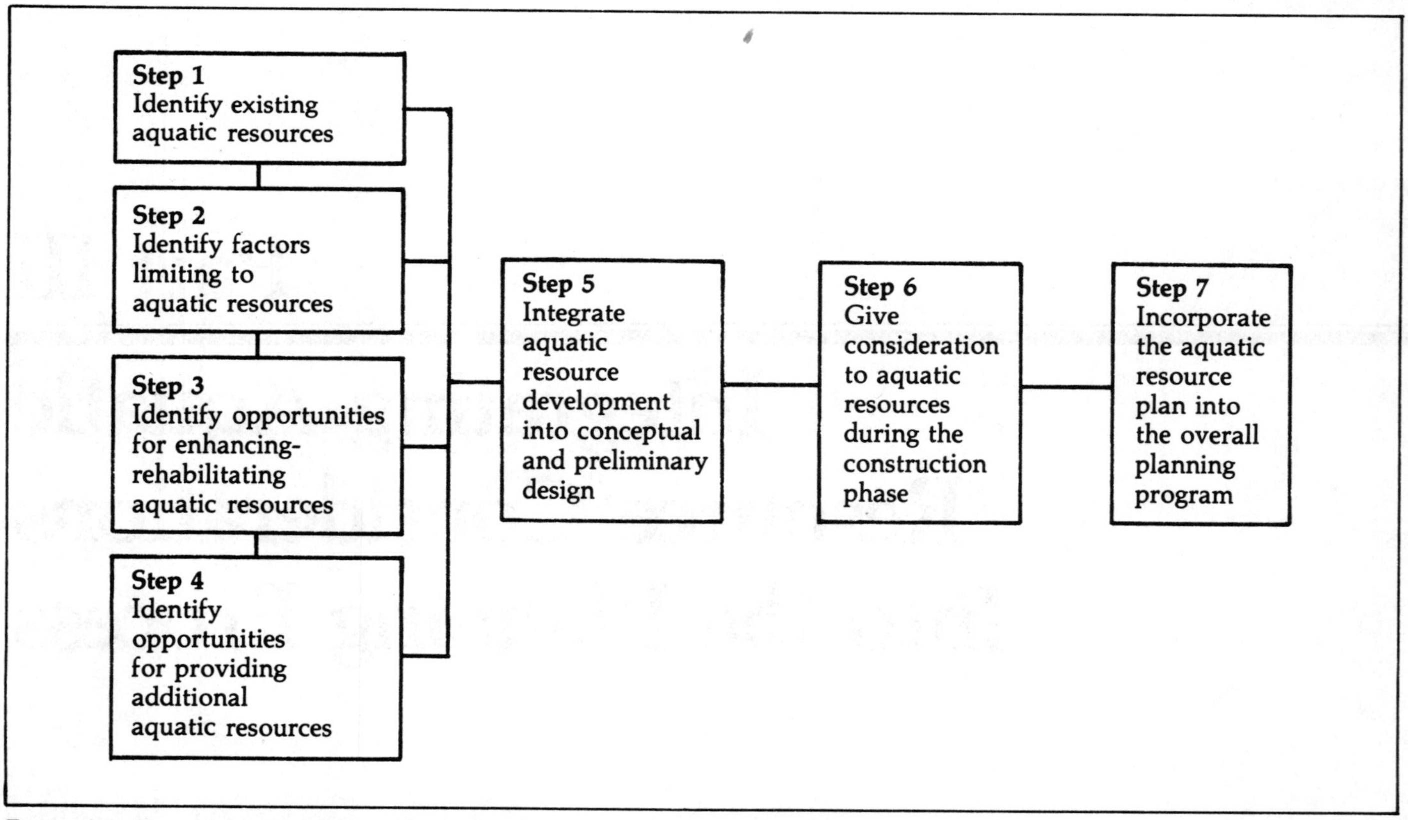

Figure 27. *Flow diagram of procedures for integrating aquatic resources into site design.*

other water bodies and surrounding land uses. A detailed inventory supplies data useful in identifying functional characteristics, outstanding features, limitations, and special management-protection measures.

The planner must recognize different water body types. For wetlands—valuable for both fish and wildlife—the U.S. Fish and Wildlife Service recommends the classification scheme outlined by Cowardin et al.[128] Because wetlands are transition areas, it is often difficult to delineate their margins; however, Lefor et al.[129] present a variety of methods for doing so. Streams and rivers are easier to define and delineate, but the inventory should be sensitive to changing physical, chemical, and biological characteristics throughout their course. Publications by Leopold and Langbein[130] and by Morisawa[131] are useful in evaluating structural and other characteristics of rivers. A U.S. Environmental Protection Agency publication on our nation's lakes details lake characteristics and man's impact on them.[132] Technical Bulletin 72 of the Urban Land Institute[133] discusses lakes and ponds together with their maintenance in urban areas. Typical inventory outlines for lakes, ponds, wetlands, streams, and rivers are listed in Appendices G, H, and I. A detailed inventory for estuarine and marine areas might include general information on longitudinal, vertical, and lateral features with data on the specific site.

(b) Evaluating functional characteristics of the aquatic resource

An inventory should also pay heed to fin and shellfish spawning and nursery areas, migratory fish passageways, fishery support areas, wildlife usage, unique biotic communities, scenic, cultural, scientific, and aesthetic values, recreation and educationally important resources, and other water uses. The inventory should reveal not only the present use, but what uses, like fishing, could be supported after a resource's restoration.

(c) Determine if critical or outstanding areas exist

Some factors to be considered in determining water bodies deserving special protection or attention are area use, relative scarcity of the aquatic resource, its proximity or accessibility for human use, and its vulnerability.

(d) Identify factors influencing the importance of resource functions

These factors may vary from an area's size and location, habitat diversity, water quality, and substrate composition with respect to fish and wildlife presence and abundance, to access and availability for recreation.

The inventory should provide a basic framework for future protection and enhancement for the site and larger-than-site planning scales. By evaluating structural and functional characteristics, critical areas and specific water quality criteria needed for enhancement and protection measures can be identified. These measures and critical area designations can be incorporated directly into performance standards and land use control schemes. Criteria should serve, too, as a basis for sound planning and decision making during the conceptual and preliminary design stages of construction.

(e) Methods of inventory and evaluation

Inventory information can be obtained from federal, state, and private sources identified in Chapter 8. Recent aerial photographs or U.S. Geological Survey topographic maps are useful in determining a water body's size, drainage patterns, surrounding land uses, proximity to other water bodies, and similar features. County planning commissions are a convenient source for information on land uses and some aquatic resource features. River basin commissions and regional planning commissions may be tapped for information. For species verification, occurrence, and distribution in a water body, or for other technical information, employ or seek the services of specialists. Maps having overlays depicting existing and proposed land uses, vegetation, erodible soils, aquifers, and steep slopes, are useful to understand resource limitations and interrelationships between land and water resources.

Step 2. Identify existing limitations to aquatic resources

Most water bodies in urban and suburban areas have been disturbed by man. In many instances, water quality is the major limiting factor, and, in others, hydrological or structural modifications limit the aquatic resource. Water bodies can be rehabilitated if corrective action is taken, thereby enhancing these resources for multiple use. It is important to identify the source of disturbance, and to develop strategies for eliminating or minimizing it. The resource inventory can be used to identify limiting influences on a body of water.

(a) Water quality considerations

Common water quality limitations include depressed oxygen levels, elevated temperatures, turbidity, nutrient levels (including phosphorous, nitrogen, and carbonaceous matter), and contamination of bottom sediments by metals and other toxic materials. Depressed oxygen levels often occur near the outfalls of sewage treatment and industrial plants. Since oxygen solubility is strongly affected by water temperature (the higher the temperature, the lower the solubility), oxygen depletion may occur locally, also, in the heated effluent waters of electric power stations or industrial plants.

Turbidity is caused by suspended organic and inorganic matter and, in some cases, by suspended microscopic plants. Sediment loading in runoff from the watershed and erosion within the stream channel are primary causes of excessive turbity, but dredging, mining activities, navigation, and recreational uses also contribute to elevated turbidity. Reasons for minimizing turbidity are discussed by

[128] L. M. Cowardin, V. Carter, F. C. Golet, and E. T. LaRoe, *Classification of Wetlands and Deepwater Habitats of the United States,* U.S. Department of the Interior, Fish and Wildlife Service, Biological Services Program, FWS/OBS-79/31, 1979, 103 pp. (Available from the Superintendent of Documents, U.S. Government Printing Office, Washington, DC 20402.)

[129] M. W. Lefor, H. H. Ridgeway, and T. B. Helfgott, *Delineation of Wetlands,* in Proceedings, Second Wetland Conference, January 9, 1974 at Storrs, CT, 1975, 118 pp.

[130] L. B. Leopold and W. B. Langbein, *A Primer on Water,* USDI, Geological Survey 1960, 60 pp. (For sale by the Superintendent of Documents, U.S. Government Printing Office, Washington, DC 20402.)

[131] M. Morisawa, *Evaluation of Natural Rivers,* USDI Water Resources Research Project No. C-1779, Final Report, 1971, 114 pp.

[132] (100) U.S. Environmental Protection Agency, *Our Nation's Lakes.*

[133] (61) J. Tourbier and R. Westmacott, *Lakes and Ponds.*

Figure 28. *Sloped natural stone riprap has been applied for streambank protection on the Mad River in Winstead, Connecticut.*

Stern and Stickle.[134] High turbidity and suspended fine-particle sediments drastically reduce the numbers and kinds of organisms present in an aquatic ecosystem.

High concentrations of dissolved nutrients encourage nuisance algae blooms in water bodies. Bloom death and decay exerts high oxygen demands resulting in fish kills on a massive scale. High concentrations of ammonia can also poison aquatic organisms. Sediments contaminated by heavy metals, pesticides, and other toxic materials have a devastating effect on aquatic species' diversity and abundance. Hence, anything that can be done to promote adequate waste treatment and control urban runoff of materials like deicing salts helps ensure water bodies suitable for fish and water-related recreation. Methods for identifying and evaluating the nature and extent of pollution and contaminants entering urban waters from streets and other sources are obtainable from the EPA. An American Fisheries Society publication[135] reviews EPA's 1976 Water Quality Criteria and describes fish response to various contaminants (**see Appendix J**).

[134]E. M. Stern and W. B. Stickle, *Effects of Turbidity and Suspended Material in Aquatic Environments—Literature Review*, Dredged Material Research Program Technical Report D-78-21, 1978, 117 pp. (Environmental Laboratory, U.S. Army Engineers Waterways Experiment Station, P.O. Box 631, Vicksburg, MS 39180.)

[135]American Fisheries Society Water Quality Section, *A Review of the EPA Red Book: Quality Criteria for Water*, R. Thurston, Russo, Fetterolf, Edsall, Barker, eds., 1979, 313 pp.

(b) Structural and hydrological alterations

Channelization, highway and residential construction, navigation, and other projects limit many aquatic ecosystems. Bridge crossings, culverts, and highway drainage can all affect river channel dynamics and the aquatic community. Channelization data are available from the Council on Environmental Quality[136], the U.S. Fish and Wildlife Service (Office of Biological Services) and from state fish and wildlife departments. Channel modification sometimes eliminates valuable inshore river shallows by dredging, filling, bulkheading, or other operations affecting streams. Channel alteration can cause major changes in stream or river erosion and sediment patterns which ultimately affect the substrate on which many aquatic life forms depend.

Dams and impoundments placed across rivers should be identified and evaluated concerning fish movement effects, downstream receiving waters' chemical quality, flow regimes, and habitat suitability for aquatic organisms. In tidal areas, note any barriers like dikes and levees across wetlands and estuaries. Also note any undersized culverts impeding or preventing sufficient fresh and salt water exchange, because changes in the salinity regime can alter the wetland/estuarine ecosystem. Many species in these areas exhibit definite salinity sensitivity.

[136]Arthur D. Little and Academy of Natural Sciences of Philadelphia, *Report on Channel Modifications*, Vols. I and II, 1973. (Prepared for the Council on Environmental Quality.)

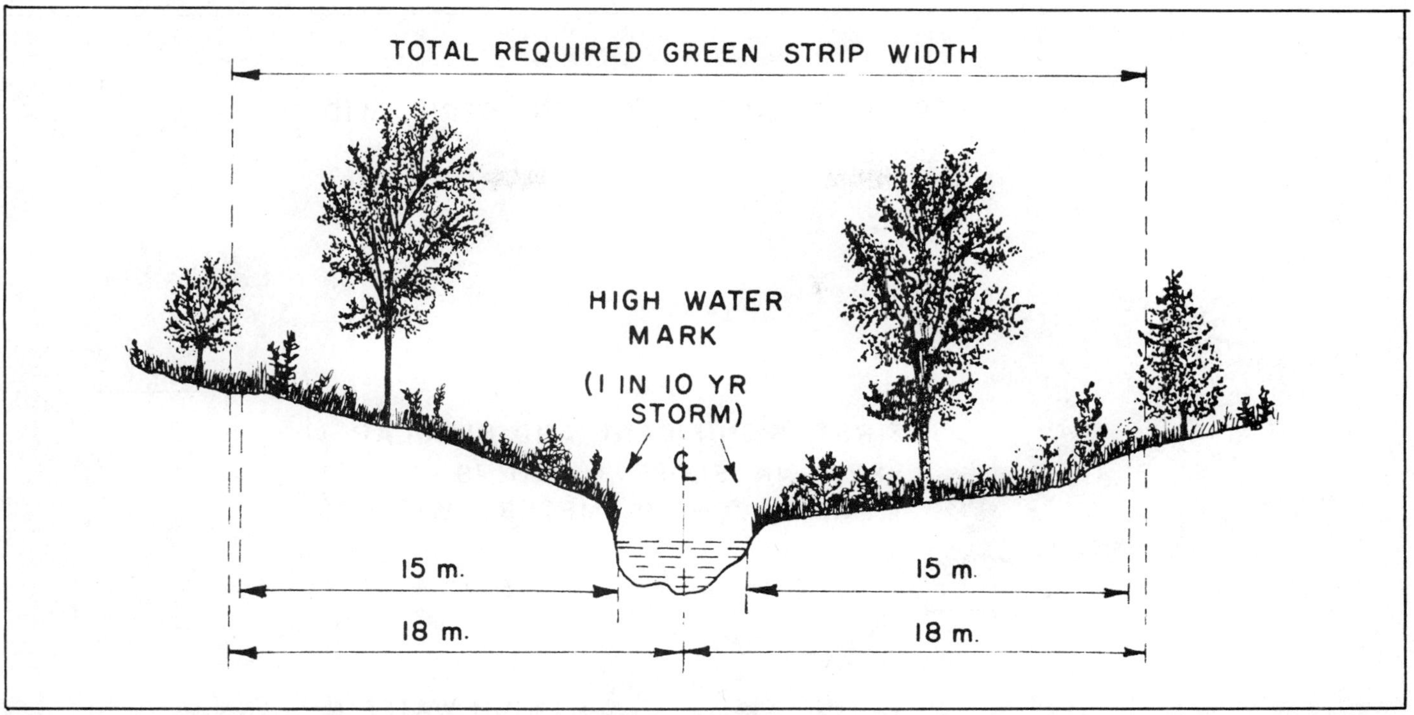

Figure 29. *A watercourse with a well-defined high water mark. In this example, 18 horizontal meters from the stream centerline is greater than 15 horizontal meters from the high water mark.*

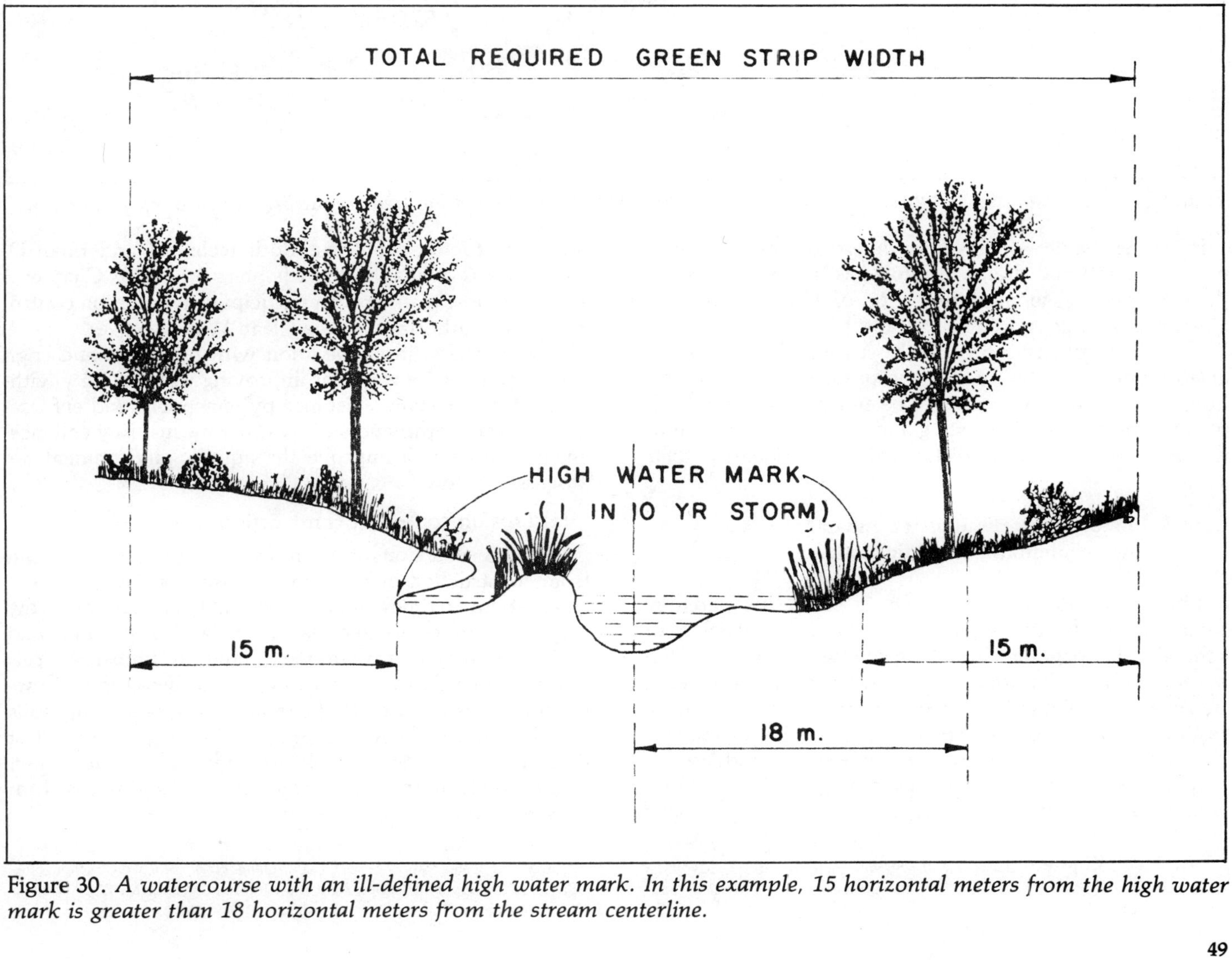

Figure 30. *A watercourse with an ill-defined high water mark. In this example, 15 horizontal meters from the high water mark is greater than 18 horizontal meters from the stream centerline.*

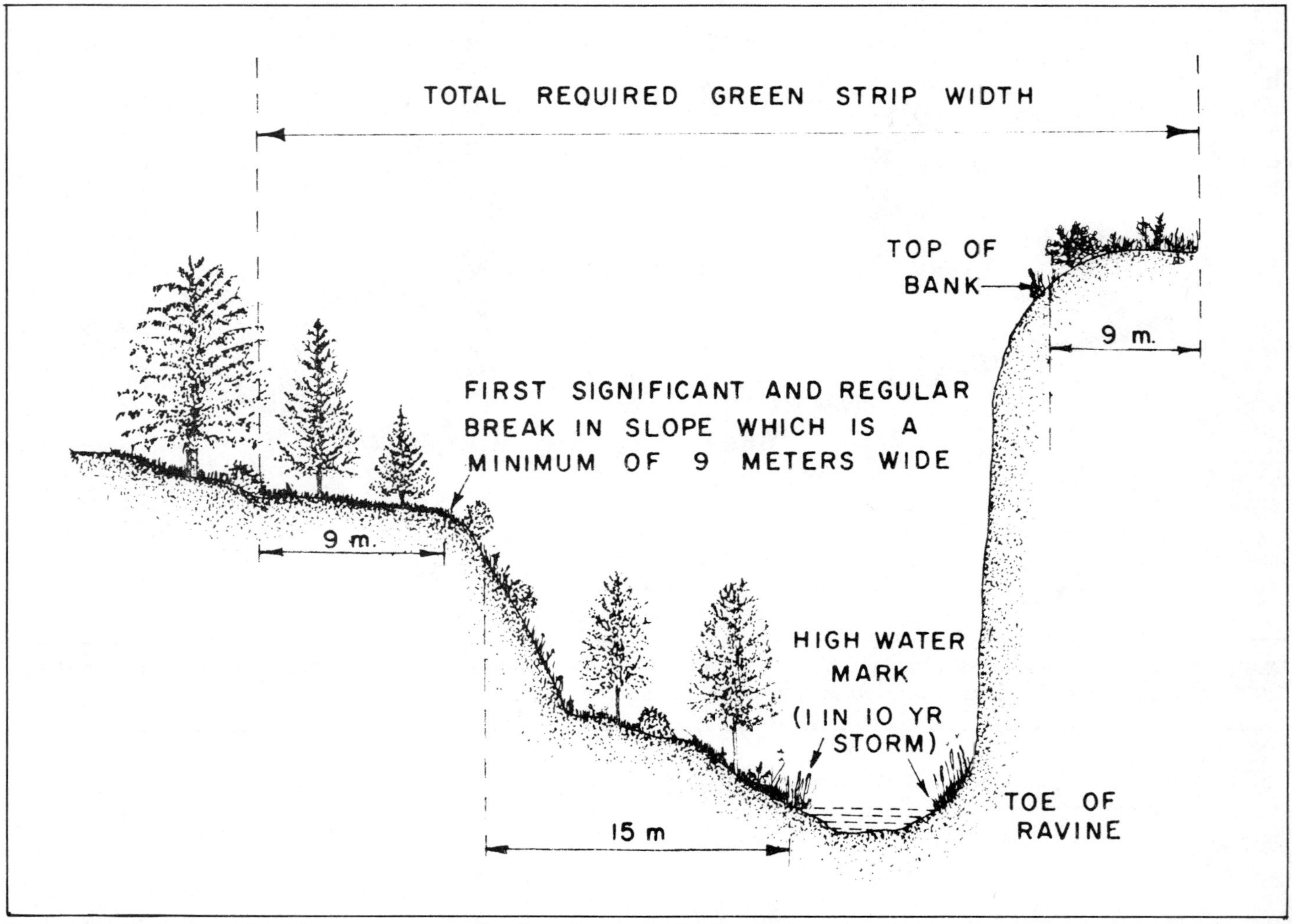

Figure 31. *Two examples of ravines or steeply sloped topography in residential areas requiring top of bank protection.*

In western states stream flow diversion into impoundments, canals, and drainage ditches is common. Loss of flow, or even seasonal modification of a river's flow can alter an aquatic ecosystem drastically.[137]

Aquatic ecosystems are highly susceptible to disturbances involving water quality and physical habitat alterations. If aquatic resources are to be enhanced or protected, planners must evaluate existing and potential resource limitations owing to these disturbances and identify their sources.

Step 3. Identify opportunities for enhancing and rehabilitating aquatic resources

Limitations identified in Step 2 should be carefully reviewed to determine opportunities for enhancing or rehabilitating degraded aquatic resources. Aquatic habitat surveys in most developed areas, and often in those yet to be developed, will indicate need for improvement. In most instances water bodies can be improved by better soil erosion and pollution control, or changes in land use on watersheds. Planners, assisted by biologists, may identify causes of aquatic habitat degradation and recommend solutions to the problems. Both technical and financial assistance is available (see Chapters 8 and 9). Chapter 5 describes how planners can participate in pollution control programs authorized by the Clean Water Act.

Site planners, in cooperation with biologists and engineers, may devise ways for improving water quality without federal or state assistance by enactment and enforcement of local ordinances or by other means. Key enhancement measures to improve the aquatic environment are discussed below.

(a) Erosion control at construction sites

The U.S. Soil Conservation Service reports that erosion from construction of highways, houses, or shopping centers is about 10 times greater than that from cultivated row crops, 200 times greater than from land in pasture, and 2,000 times greater than from land in timber.[138] Yet, erosion and sedimentation can be controlled if the following principles are used in the treatment of land-using soils suited for development: leave the soil bare for the shortest time possible; reduce the velocity and control runoff flow; detain runoff on the site to trap sediment; and release runoff safely to downstream areas.

[137]J. C. Fraser, *Regulated Discharge and the Stream Environment*, pp. 203-205 in River Ecology and Man, R. T. Oglesby, C. A. Carlson, and J. A. McCann, eds., Academic Press, NY, 1972.

[138]USDA Soil Conservation Service, *Controlling Erosion on Construction Sites*, USDA Soil Conservation Service, Agriculture Information Bulletin 347, 1970, 32 pp. (U.S. Government Printing Office, Washington, DC 20402.)

(b) Stream bank stabilization and protection

Erosion from bank instability of rivers and other water bodies and within river channels themselves can be reduced by stabilization through vegetation and by artificial measures suggested by Tourbier and Westmacott.[139] However, vertical-walled structures like sheet pilings and bulkheads are less well suited for aquatic life; sloped natural stone riprap construction is preferable.

One of the most effective means for enhancing aquatic resources is to maintain vegetative buffers along water bodies and by developing such buffers where they do not occur. When composed of grasses, shrubs, and trees, buffers not only help stabilize stream banks, but entrap sediment, aid in adsorbing pollutants in overland runoff, and provide food and cover for fish and wildlife. Soil Conservation Service offices can provide plant listings suitable for different localities. Also, six regional manuals published in 1977 and 1978 by the Office of Biological Services, U.S. Fish and Wildlife Service give information about shrub and tree species tolerance to water level changes in riparian and wetland communities. Buffer strip width needed for effective erosion control depends on several factors, including percent of slope. On municipal watersheds greater width than usual is needed.[140] The Department of Fisheries and Oceans Canada[141] has developed a useful guide for minimum buffer strip width recommended to protect aquatic life. This guide is based on adjacent slope characteristics and water course definition. (See Figures 29, 30 and 31.) Generally, a border extending at least 15 meters from the high water mark with a minimum of 36 meters total is recommended.

(c) Removal of sediment

Where massive amounts of sediment have accumulated and are known to be a major factor in limiting the health of an aquatic ecosystem or the use of the water for water-based recreation, excessive sediment can be removed. It should be recognized, however, that this is a temporary means for enhancing lakes and ponds because they will fill in again quickly if accelerated erosion on the watershed is not prevented. Unless care is used in the process, sediment removal operations can contribute to turbidity; dissolved oxygen levels can be reduced through oxidation of organic and other sediment-bound substances; nutrients, metals, and other toxic substances can be released from stirred-up sediments, altering water quality. Therefore, suction dredges or other methods for sediment removal reducing re-suspension to a minimum should be used. Information about dredging methods, its undesirable effects, disposal of dredged materials on aquatic organisms, and alternative dredging techniques is available.[142]

(d) Rehabilitation of lakes

Excess sediment removal from lakes is but one of many methods available for rehabilitating lakes and reservoirs. Some of these were discussed in Chapter 5. A useful review of lake rehabilitation techniques is provided by Dunst et al.[143] and means for improving western reservoirs are suggested by Nelson et al.[144] A common objective of lake or reservoir improvement is the control of nuisance growths of aquatic plants interfering with recreation and which have other undesirable effects. Aquatic plant control information is available from the U.S. Fish and Wildlife Service, U.S. Army Corps of Engineers, U.S. Water and Power Resources Service, and from state agencies.[145] [146] Planners and municipal decision makers are reminded that excessive nuisance plant growth removal will only be temporary if they exert no control over nutrient quantities entering the lakes.

(e) Rehabilitation of rivers and streams

In addition to sediment removal, stream rehabilitation may involve removing channel obstructions or installing fish ladders and the like to permit fish passage. Debris dams and other obstructions, while often considered to be a major factor in fish migration, flooding, or eliminating spawning areas, sometimes should be left intact because they serve as important habitats for aquatic life. Channelized streams can often be improved by installing stone or log deflectors, gabions, and low dams. Many urban streams and rivers may be improved insofar as aquatic resources are concerned by restoring inshore shallows via removing hydrological obstacles like revetments. Where bulkheads are in disrepair, they can be removed and replaced by sloping riprap to offer a more hospitable habitat than vertical walled structures.

(f) Wetlands restoration

Although dikes may be necessary to manage water levels in some wetlands effectively, many of the dikes are in disrepair or obsolete. Often their removal can rejuvenate a wetland system and surrounding water bodies dependent on wetland functions. By restoring tidal flushing in coastal wetlands, deteriorated water quality caused by sediment and toxic substance accumulations can be reduced. Simultaneously, marsh productivity and usefulness as shelter, fish and waterfowl feeding, and a fish spawning ground will be increased. By furnishing connections between wetlands, corridors valuable for fish movement are created. Water circulation can be improved, too, by replacing undersized culverts with ones that can better accept runoff and assure sufficient tidal flow.

(g) Removal of undesirable species

In addition to controlling undesirable growths of vegetation, control of fish species such as carp may be warranted

[139] J. Tourbier and R. Westmacott, *Water Resources Protection Measures in Land Development. A Handbook,* Water Resources Center, University of Delaware, Newark, DE, 1974, 237 pp.

[140] G. R. Trimble and C. S. Sartz, *How Far From a Stream Should a Logging Road be Located?,* pp. 339-341 in Journal of Forestry, (May 1957), 1957.

[141] Department of Fisheries and Oceans Canada, *Guidelines for Land Development and Protection of the Aquatic Environment,* Fisheries and Oceans Canada, Pacific Region, 1090 W. Pender Street, Vancouver, British Columbia, V6E 2PL.

[142] N. D. Hirsh and L. H. DiSalvo, *Effects of Dredging and Disposal on Aquatic Organisms,* Dredged Materials Research Program, Technical Report D6-78-5, 1978, 41 pp. (Environmental Laboratory, U.S. Army Engineers Waterways Experiment Station, P.O. Box 631, Vicksburg, MS 39180.)

[143] (102) R.C. Dunst, S.M. Born, P.D. Uttormark, et al., *Survey of Lake Rehabilitation Techniques and Experiences.*

[144] (103) R. W. Nelson, G. C. Horak, and J. E. Olson, *Western Reservoir and Stream Habitat Improvements Handbook.*

[145] S. A. Nichols, *Mechanical and Habitat Manipulation for Aquatic Plant Management: A Review of Techniques,* Technical Bulletin No. 7, 1977, 34 pp. (Department of Natural Resources, Madison, WI.)

[146] L. A. Lueschow, *Biology and Control of Selected Aquatic Nuisances in Recreational Waters,* Technical Bulletin No. 57, 1972. (Department of Natural Resources, Madison, WI.)

in some situations because they contribute to high turbidity levels with consequent effects on a water system's productivity. Information on fish control is available from state fish and wildlife departments and the U.S. Fish and Wildlife Service (see Chapter 8 and appendices). However, carp can offer a type of fishing acceptable to many urban residents.

Step 4. Identify opportunities for providing additional aquatic resources

In some settings new water areas can be created to provide additional recreational opportunities and, in some cases, serve as a trade-off for other aquatic resources that have been disrupted or destroyed. For example, in Coos Bay, Oregon, efforts are under way to restore and create additional wetlands where filling and dredging of intertidal lands has occurred.[147] Treatment effluents may be useful for creating new aquatic areas as well. Siting and design considerations are important, whether the water area to be developed is a wetland, pond, lake, or other water body, even when it is intended specifically for recreation or multiple use.

(a) Ponds and reservoirs

Ponds or reservoirs created by damming streams or drainages often require expensive diversion structures and large scale dams to accommodate watershed runoff and, in large reservoirs, wind-generated waves can cause erosion. Therefore, many siting and engineering design aspects must be considered to ensure a valuable aquatic resource following construction. Useful guides are available from state conservation departments,[148] from the U.S. Department of Agriculture,[149] [150] [151] and from other organizations like the Urban Land Institute.[152] Among factors to be considered are the amount and quality of runoff and influent waters,[153] size and configuration of the impoundment, need for water level control devices (see Part II of guide), and downstream impoundments effects created by river damming (see Chapter 3).

(b) Wetlands

Most wetland establishment schemes involve dredge disposal sites, extant sea level habitats such as tidal flats, or excavating depressions in upland habitat into which water is introduced. The U.S. Fish and Wildlife Service has had considerable experience in managing and creating wetlands for waterfowl, e.g., at the Ottawa National Wildlife Refuge near Toledo, Ohio. Perceptive information about planned marsh establishment work in the United States is provided by Garbisch.[154] Wetland rehabilitation usually involves water level and vegetation control along with occasional seeding or transplanting of aquatic plants. Lewis et al.[155] have described efforts to mitigate wetland habitat losses associated with expansion of a shipyard in Tampa Bay, Florida by creating mangrove-marsh habitat on dredged disposal islands. Though newly created and rehabilitated wetlands differ from those that have existed for thousands of years,[156] the continual organic matter build-up in the substrate of newly vegetated wetland affords new opportunities for colonizing and establishing benthic organisms and habitats for fish and wildlife as wetlands mature.

(c) Ditches and canals

Drainage ditches have reduced wetlands in both urban and rural areas. While more drainage is not encouraged, present roadside ditches establish aquatic habitats in areas where they are lacking. Existing ditches might favor aquatic organisms by maintaining vegetated strips along them.

Canals, prominent features of new towns in the IJsselmeerpolders of the Netherlands, are important for recreation.[157] Water area diversity and extent is achieved by widening canal portions to form a lake, varying the canal bank's slope, and diversifying the vegetation, to encourage different fish and wildlife populations.

Step 5. Integrate acquatic resource development into conceptual and preliminary design

A comprehensive planning format needs an integrated approach to land development and aquatic resource control. Municipalities' and developers' initiative and interest in integrating aquatic resource development into conceptual and preliminary design phases of a land development project can help preserve and enhance the resource during site construction. Planning and design guidelines for accomplishing this follow.

(a) Siting considerations

Preliminary evaluation of a site's suitability for a particular land use or construction activity should be based on an inventory of key natural resource features.[158] Features to be considered are:

(i) Base soil limitations and suitabilities on erodibility, water filtration capacity, on geology, slope length and

[147]E. T. LaRoe, *Mitigation: A Concept in Wetland Restoration*, pp. 221-224 in Proceedings of the National Wetland Protection Symposium, 1977. (U.S. Fish and Wildlife Service.)

[148]J. G. Dillard, *Missouri Pond Handbook*, Undated, 60 pp. (Missouri Department of Conservation, P.O. Box 180, Jefferson City, MO 65101).

[149](62) USDA Soil Conservation Service, *Ponds for Water Supply and Recreation.*

[150]U.S. Department of Agriculture, *Warm-Water Fish Ponds*, Farmers Bulletin No. 2250, 1977, 14 pp.

[151]U.S. Department of Agriculture, *Building a Pond*, Farmers Bulletin No. 2256, 1973, 13 pp.

[152](61) J. Tourbier and R. Westmacott, *Lakes and Ponds.*

[153]N. A. Whalen, *Nonpoint Source Control Guidance Hydrological Modifications*, 1977. (U.S. Environmental Protection Agency, Water Planning Division, Nonpoint Sources Branch, Washington, DC 20460.)

[154]E. W. Garbisch, *Recent and Planned Marsh Establishment Work Throughout the Contiguous U.S.: A Survey and Basic Guideline*, U.S. Army Engineers Waterway Experiment Station Contract Report D-77-3, 1977, 41 pp. (U.S. Army Corps of Engineers, U.S. Waterways Experiment Station, Vicksburg, MS.)

[155]R. R. Lewis, C. S. Lewis, W. K. Fehring, and J. A. Rodgers, Jr., *Coastal Habitat Mitigation in Tampa Bay, Florida*, pp. 136-40 in The Mitigation Symposium—A National Workshop on Mitigating Losses of Fish and Wildlife Habitats, 1979. (Rocky Mountain Forest and Range Experiment Station, USDA Forest Service, Fort Collins, CO.)

[156]E. W. Garbisch, *Wetland Rehabilitation*, pp. 217-219 in Proceedings of the National Wetland Protection Symposium, 1977. (USDI Fish and Wildlife Service.)

[157]J. A. van den Berg, A. K. Constandse, R. W. Greiner, J. deJong, M. Loenen, K. Rijnierse, and E. Schultz, *Water in New Towns in the IJsselmeerpolders*, Rijksdienst voor de IJsselmeerpolders Smedinghuis, Lelystad, Flevobericht, nr. 151, 1979, 53 pp.

[158]McHarg, *Design with Nature*, 1969, 198 pp. (Doubleday, Natural History Press, Garden City, NY.)

gradient. Avoid development in those areas prone to nonpoint source loading. Areas having steep slopes (greater than 15%) and short slope length or where soils are highly erodible should be excluded from development. Maintain them instead for permanent open space because their development is likely to affect aquatic ecosystems adversely, requiring expensive sediment and stormwater control.

(ii) Consider critical aquatic resources or systems highly vulnerable to disturbance, like wetlands harboring threatened or endangered species of fish and wildlife. Identify areas overlying underground aquifers. Rivers—especially those included in state or federal scenic river lists, floodplains, and areas representing important ecological, educational, and recreational opportunities should be examined.

(iii) Note hydrologic and hydraulic features (surface and subsurface). Sites where development would impact on water bodies having hydrologic significance (e.g., water supply, flood control) should be excluded from development.

(iv) Perceive any watershed problems upstream or downstream from proposed construction sites. Unless methods surely control disturbances from proposed construction, it is best to avoid areas where such impacts exacerbate drainage problems.

Vegetative buffer zone installation along water courses and exclusion of sewage disposal systems proximal to their margins need identification and control during preliminary design phases, thus preventing future adverse environmental impact. Where water bodies support cold water fisheries, it is important to identify and plan for preservation or development such that wooded areas and springs supplying cool water will continue to do so.

With respect to highway routing and construction designs, any streams used by anadromous fish whose migration might be prevented by culverts of improper design or size should be noted. Guides for solving these problems are available.[159] [160] Under guidance from biologists, borrow pit shapes can be altered slightly during excavation to make them more suitable for fish, waterfowl, and other wildlife. Some suggestions on how to make such water areas suitable for fish are offered by Moulton.[161]

Airport planning in relation to fish, wildlife, and aquatic sites to avoid accidents like aircraft-bird collision is considered in a companion volume.[162]

Power plants must be carefully sited to avoid disturbing habitats for threatened and endangered species. Research has resulted in screening devices at cooling water intakes which now help prevent unnecessary aquatic animal mortality through screen impingement, and stop entrainment in water going through the plant. Work on design, alignment, and configuration of power lines and towers to prevent bird loss through collisions or electrocution is going forward. It has been suggested that new line construction over areas where large waterfowl concentrations occur be avoided.[163]

Building density and watershed impenetrability in developed areas in relation to water body impairment is another important consideration in preliminary designs. In his studies of Piedmont streams in Maryland, Klein[164] stated that for sensitive stream ecosystems like those supporting trout populations, watershed imperviousness should not exceed 10 percent. He developed a table showing the maximum amount of watershed that can be developed based on 10 and 15 percent imperviousness, and according to land use categories varying from 1-acre home lots to shopping centers. Leopold[165] gave similar information about lot sizes in residential areas and on the basis of an area's percentage served by storm sewers.

(b) Planning for water pollution and stormwater control

Once a suitable site has been determined, an effective water pollution abatement plan should be devised to protect aquatic resources both at the site and within the watershed. The plan should reflect planning principles with regard to the landscape unit addressed above and, relative to water resource protection, should consider the following: groundwater disturbance; construction on or near potential landslide or mudslide area; stream crossing structures; land fill, culvert, dike, and building encroachment on stream flow; influences on stormwater runoff imposed by an increased area of impervious streets, parking lots, and buildings; changes in drainage caused by diversions and gradings, sediment spoil and other solid wastes disposal; floodplain excavation work; stream channel modification; petroleum waste, pesticide, and other chemical disposal; control of dust; access and haul road construction; sewage treatment; construction site proximity to streams, lakes, and other vulnerable water bodies; vegetation alteration; wetland modification; natural circulation pattern interference with respect to tide; and sediment erosion. Methods for minimizing and controlling potential water quality impacts associated with these activities are discussed by Tourbier and Westmacott,[166] Department of Fisheries and Oceans Canada,[167] Carroll[168] and by the U.S. Environmental Protection Agency.

A plan for sediment and stormwater control, which also relates to pollution control, should be devised for all development sites. This plan should include means for:

[159](78) USDA Forest Service, *Wildlife Habitat Improvement Handbook.*

[160]W. A. Evans and F. B. Johnston, *Fish Migration and Fish Passage: A Practical Guide to Solving Fish Passage Problems,* U.S. Dept. of Agriculture, Forest Service, Region 5, 1974.

[161]J. C. Moulton, *The Fishery Potential of Four Aquatic Environments Created by Interstate 91 Construction in Massachusetts,* M.Sc. Thesis (University of Massachusetts, Amherst, 1970), 86 pp.

[162]D. L. Leedy, R. M. Maestro, and T. M. Franklin, *Planning for Wildlife in Cities and Suburbs,* USDI Fish and Wildlife Service, FWS/OBS-77/66, 1978, 64 pp.

[163]W. L. Anderson, S. S. Hurley, and J. W. Seets, *Waterfowl Studies at Lake Sangchris,* Final Report, 1975, 15 pp. (Illinois Natural History Survey, Urbana, IL.)

[164](13) R. D. Klein, *Urbanization and Stream Quality Impairment.*

[165]L. B. Leopold, *Hydrology for Urban Land Planning—A Guidebook on the Hydrological Effects of Urban Land Use,* Geological Survey Circular 554, USDI Geological Survey, 1968, 18 pp. (U.S. Government Printing Office, Washington, DC 20402.)

[166](61) J. Tourbier and R. Westmacott, *Lakes and Ponds.*

[167](141) Department of Fisheries and Oceans Canada, *Guidelines for Land Development and Protection of the Aquatic Environment.*

[168]Allen Carroll, *Developer's Handbook,* Undated, 60 pp. (State of Connecticut, Department of Environmental Protection, State Office Building, 165 Capitol Avenue, Hartford, CT 06115).

(i) controlling water runoff speed and volume on the construction site—

Options for accomplishing this objective include natural drainage systems,[169] grassed swales instead of concrete ditches along roadsides in certain types of housing developments,[170] diversionary structures to delay runoff delivery to watercourses, rooftop or parking lot ponding or use of recreation areas for temporary water storage, lawn and golf course aeration increasing infiltration, inline storage in sewers, instream, side-channel and off-channel water course storage, and use of dry detention basins or small permanent sediment ponds.

Sediment basins and stormwater control ponds can be valuable for fish and wildlife if allowed to remain in place after construction. Design criteria for multiple use ponds are being developed in Howard County, Maryland by the Department of Public Works, Soil Conservation Service, and the Urban Wildlife Research Center, Inc.

(ii) minimizing pollutant loadings—

Urban street routing and design can govern pollutant quantities entering urban water bodies. Polluted water from urban streets contains toxic contaminants from many sources, including automobiles and lawns where insecticides have been used. Serious damage to aquatic organisms may occur as a result, particularly during initial pollution loadings resulting from storms. A study by Sartor and Boyd[171] showed that the quantity of contaminant material entering receiving waters averaged about 1,400 pounds per curb mile. Some EPA suggestions for reducing nutrients, toxic, and oxygen-demanding substances introduced from street runoff during a storm are:

- select roadway sites minimizing the area draining directly into the receiving water body;
- use low curbs where the road joins flat unpaved areas or those sloping gently away from the street surface—this will ease dust and dirt deposition into grass and gravel and reduce the deposition rate of runoff water;
- consider using porous pavement where climate and soil types will permit it;
- intensify and improve street-cleaning operations to reduce urban runoff effects;
- design curbs and gutters to ease concentration and collection of particulate material.[172]

[169]D. A. Rickert and A. M. Spieker, *Real-Estate Lakes: Water in the Urban Environment,* Circular 601-G, 1971, 19 pp. (U.S. Geological Survey, Reston, VA 22092.)

[170]Northern Virginia Planning District Commission, *Guidebook for Screening Urban Nonpoint Pollution Management Strategies,* Final report prepared for Metropolitan Washington Council of Governments, 1979, 121 pp. (Northern Virginia Planning District Commission, 7309 Arlington Blvd., Falls Church, VA 22042.)

[171]J. D. Sartor and G. B. Boyd, *Water Pollution Aspects of Street Surface Contaminants,* U.S. Environmental Protection Agency, EPA-R2-72-081, 1972, 236 pp. (U.S. Government Printing Office, Washington, DC 20402.)

[172](70) D. G. Shaheen, *Contributions of Urban Roadway Usage to Water Pollution.*

Where possible, routing runoff water through a wetlands area before it enters a river or lake helps remove sediment and suspended solids.

(iii) determining sediment control measures at the site—

A variety of sediment control measures are considered in the next step (6). However, careful consideration should be given in the preliminary design stage to ensure cost-effective and environmentally sound measures for entrapping sediment at the development site. Helpful information and guides are available from the U.S. Environmental Protection Agency and other sources.[173] [174]

(iv) determining soil stabilization practices in relation to seasons—

When possible, construction should be scheduled to avoid heavy rainfall months so prompt reseeding and sodding can reduce erosive soils exposure.

(c) Planning tools

Effective site plans are built primarily by compiling and analyzing data derived from local, county, state, and federal records, supplemented as necessary by special studies or surveys. Suggested information sources are presented in Chapter 8. Maps constitute a valuable tool. Base maps having a scale of 2,000 feet to the inch (contour intervals = 20 feet) are useful for generalized site planning purposes, but scales ranging from 50 feet to the inch (contour interval = 5 feet) are more suitable for average-sized sites, stormwater, and sediment control plans.

Map and other data availability does not eliminate the need for detailed on-site studies by specialists to provide guidance on identifying critical areas and designing stormwater, sediment, and erosion control methods. Depending on the development site, agronomists, soil scientists, geologists, hydrologists, engineers, landscape architects, economists, site planners, and biologists might participate. Composite maps depicting specific natural resource limitations as they relate to aquatic resource protection are particularly valuable during preliminary design stages to identify optimal development sites and areas having moderate to severe restrictions on development. Detailed methods for preparing individual resource and composite maps are given by McHarg.[175]

Step 6. Give consideration to aquatic resources during the construction phase

Although judicious development site selection and layout design avoids many erosion-sedimentation and pollution problems, controls must be instituted during the construction phase for any development project. Some of these have been alluded to in the previous section; others will be addressed here.

(a) Staging of construction

Construction projects and associated grading and revegetation operations should be staged so that a

[173](61) J. Tourbier and R. Westmacott, *Lakes and Ponds.*

[174]ULI, ASCE, NAHB, *Residential Erosion and Sediment Control: Objectives, Principles and Design Considerations,* 1978, 63 pp. (The Urban Land Institute, Washington, D.C. 20005 and American Society of Civil Engineers, New York, NY 10017.)

[175](158) McHarg, *Design with Nature.*

Figure 32. *Fishermen enjoy the Delta-Mendota Canal a few miles south of Firebaugh, California. This is part of a Water and Power Resources Service endeavor called the Central Valley Project.*

minimum amount of soil surface is exposed at any time. For large scale developments that take some time to complete, it is not necessary to denude and grade the entire tract at one time. Schedule earth moving operations at seasons when seeding or planting can be done to revegetate the disturbed areas quickly.

(b) Erosion control

Surface roughening involving scarification and serration of exposed slopes reduces runoff velocity and thereby the extent of erosion. With slopes roughened, seed and fertilizers are less apt to wash out and vegetation can become better established. Various methods for intercepting and diverting runoff and establishing vegetation have already been described. Many local SCS offices can provide listings of plants suitable for specific sites, based on slopes, soil condition, and maintenance expectancy. They can provide information, too, on structural methods for stabilizing soils.

In building siting, and construction, natural vegetation, including shrubs and ground cover, should be left intact insofar as possible. Trees provide shade and protection from wind, add to residential property value, attract wildlife, and reduce erosion. The Agricultural Research Service has developed criteria for determining whether trees are valuable enough to justify removal from construction sites for transplanting elsewhere, and guides to methods that can be used to protect them during construction.[176] Detailed designs and associated costs for erosion control methods are available.[177] [178]

(c) Control of sedimentation

Sediment abatement controls for a long term project's life should be instituted during the construction phase, also. Many of the cited references apply, and, as indicated previously, maintaining vegetation buffer strips along waterways is a useful approach. Structural controls, including gravel inlet filters, sediment traps, permanent wet sediment basins, and other sediment control practices can be incorporated directly into the stormwater management plan for a construction site.

[176]U.S. Department of Agriculture, *Protecting Shade Trees During Home Construction*, Agricultural Research Service, Agriculture Research Service, Agriculture Information Bulletin 347, 1970, 32 pp. (U.S. Government Printing Office, Washington, DC 20402.)

[177](174) ULI, ASCE, NAHB, *Residential Erosion and Sediment Control: Objectives, Principles and Design Considerations.*

[178]R. E. Thronson, *Comparative Costs of Erosion and Sediment Control, Construction Activities*, EPA-430/9-73-016, 1973, 205 pp. (U.S. Environmental Protection Agency, Office of Water Program Operations, Water Quality and Non-Point Source Division, Washington, DC 20460.)

(d) Other pollutants

Many pollutants are kept from escaping the construction site by effective stormwater and erosion-sediment controls and good "housekeeping" practices in disposing of excess paints, asphalt products, pesticide containers, etc.

(e) Hydrological and in-channel modification

Hydrological and in-channel modification during the construction phase should be kept minimal. Riparian vegetation should be protected during development, and heavy equipment should be kept out of streams. To stabilize shore lines, use native vegetation or gabions rather than impervious or vertical wall structures like bulkheads. If bulkheads are used, they should be located no farther waterward than mean high tide. Sloping riprap construction, using appropriately sized rock, is preferable to concrete structures.

During some construction projects, obstructions and debris like fallen trees or limbs are often removed because they are believed to contribute to flooding or present an "untidy" appearance. Unless such debris and obstructions are a major factor in flooding or interfere with recreation, they should be left within the channel because they serve as important habitats for aquatic organisms.

Consideration should be given, also, to the potential effects of development on the hydrology of adjacent or nearby water bodies. Projects should be designed to ensure natural circulation patterns, salinity regimes, and nutrient distributions so that aquatic life is not altered in wetlands, estuaries, and marine settings. Thus, docks and piers should be designed so they do not restrict circulation and should be located in areas where there are existing channels or where initial and maintenance dredging will be minimal. In tidal areas where wetlands can be affected by development, culverts should be large enough to accommodate tidal flow along streams.

Step 7. Incorporate the aquatic resource plan into the overall planning program

Aquatic resources planning is one important component of a larger planning process in which many other socioeconomic, legal and natural resource elements must be considered. Before implementing a water resources plan, it should be reviewed to see how well it fits into the overall plan. Sound land use planning principles involve comprehensive consideration of an area's natural resources and the restrictions or limitations they place on certain uses, including construction. The aquatic resource is an integral part of this larger interacting system. Land management practices recognizing suitable land uses and sound pollution control methods and standards based on natural constraints, protect water bodies found in that system. Even in areas where there are water shortages and many competing uses for water, there are opportunities for enhancing and protecting aquatic ecosystems. In reviewing the water resources plan and integrating it into the larger one, additional opportunities for remodeling urban fishing and waterfront recreation may be identified; if so, they can be incorporated in the plan.

Photo: Heritage Conservation & Recreation Service

Figure 33. *Culverts should be designed large enough to permit fish passage.*

Chapter 7

Special Considerations at the Municipal and Regional Levels

This chapter concerns planning typically undertaken by municipal planning departments, county and regional planning commissions, watershed associations, conservation commissions, state, and interstate groups. Municipal planning activities are emphasized because greater implementation can be achieved here than at state or interstate levels.

In regional planning, aquatic resource considerations focus mainly on the preservation and maintenance of continuous greenbelts based on water corridors. A major large-scale planning function should identify key aquatic areas to be kept free of any land development. Steps below on integrating aquatic resource concerns into regional and municipal planning follow generally, but with some modification, principles and approaches suggested for site planning.

Step 1. Identify existing aquatic resources and determine their relative values

One of the most important regional planning functions with respect to aquatic resources is the identification and

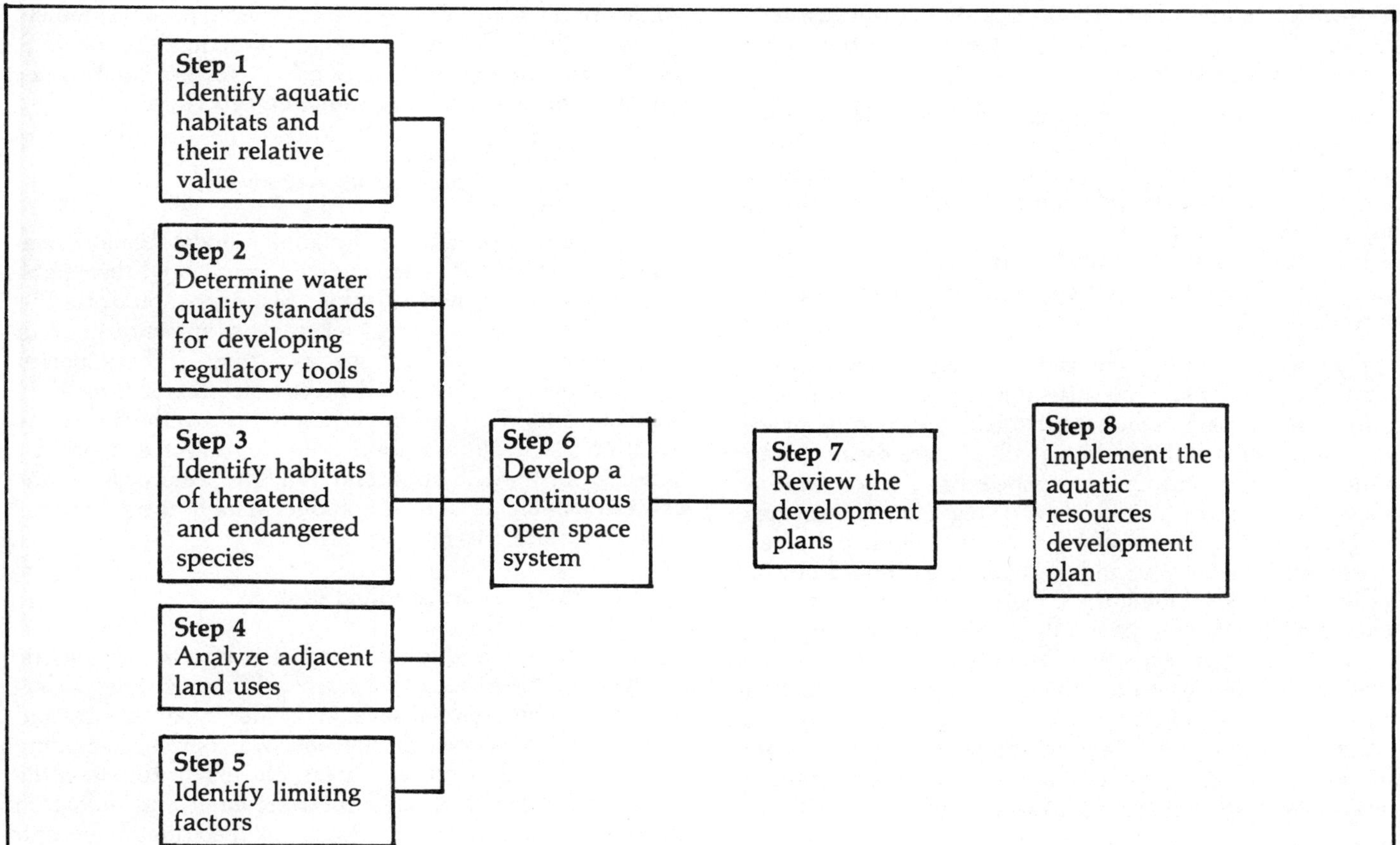

Figure 34. *Flow diagram of basic aquatic resource planning procedures at regional and municipal levels*

preservation of important habitats. Maintaining habitat diversity is a prime consideration in preserving ecological stability. To maximize diversity, limited or unique habitats must be preserved or protected. Where these are occupied by threatened or endangered species, this is a legal requirement. In this step, aquatic areas having water-based recreation, aesthetic, and public education value should also be identified.

Detailed site level inventory data is unnecessary at the broad regional level. However, information about the extent and distribution of water bodies, water quality, biological productivity, and the type and quality of fisheries present should be obtained. Much helpful information can be noted in regional inventories about integrating fish, wildlife, and recreation elements into the plan. Topics may include:

- presence of cold water fisheries in an area of typically warm water fisheries
- shellfisheries and seeding areas
- watercourses used as spawning, nursery, or breeding areas by finfish
- water bodies and wetlands known to be important to waterfowl or other wildlife
- free-flowing streams where intermitent or channelized streams predominate
- water bodies potentially useful for urban/suburban fishing, canoeing, boating, or other recreation.

Information for this planning level can be obtained through reconnaissance surveys, river basin commissions, local authorities, and state agencies (see Chapter 8).

A more detailed inventory is required at the municipal or township level. Many townships in the northeastern U.S. can already offer such inventories. The data base developed can serve as:

- master plan update and revision sensitive to the natural environment
- identification and protection of valuable natural resources for ecological, educational, scientific, aesthetic, and recreational purposes.
- a basis for discovering cause for pollution
- a platform for performance standards or other regulatory controls.

Inventories like this are typically limited to water's chemical characteristics because aquatic resource concerns at this level focus mostly on eutrophication and its problems. However, the more data available, the easier it is to define causal relationships, prescribe rehabilitation and enhancement measures, and defend regulatory measures developed for aquatic resources. At the municipal level, for example, gravel pits and rock quarries should not be overlooked in developing the inventory. Gravel mining in some areas nearly always results in lakes which can often provide housing or recreation development settings incorporating fish and wildlife habitats. Because construction minerals are scarce in many areas, some local governing units pass ordinances to assure gravel extraction before surface development proceeds, and set forth conditions for gravel extraction and mined area restoration.[179]

[179]Virginia (Commonwealth of), Fairfax County, *A Natural Resources Development Plan*, Fairfax County Planning Division, Master Plan Section, 1961, 45 pp.

Step 2. Determine water quality standards for development of performance standards or other regulatory tools

Aquatic system inventory data are useful to municipalities for developing performance standards or regulatory tools permitting growth while preserving current amenities for residents. Any regulatory control must be based upon some standard. An appropriate one for water quality considers "healthy" a water quality which presently exists within the municipality, i.e., not downstream from a sewage treatment facility. Such purity represents water quality the municipality would like to maintain, even after development.

"Healthy" water quality determinations can be made using inventory data developed as previously described. Water quality meeting or exceeding these standards indicates a "healthy" surface water for the municipality.

Existing standards usually prescribe minimum acceptable conditions for a specified use, i.e., those conditions not meeting such standards might adversely affect aquatic organisms.

Once water body conditions within a municipality are known and "healthy" standards defined, one can determine if water quality is deteriorating. If it is, the cause (source) should be established as a basis for intelligent, environmentally sound land use planning. Note that "healthy" or acceptable standards developed in this manner apply generally to relatively undeveloped parts of the municipality. Specificity achieved depends on the number of sampling stations used and, therefore, may eliminate pre-and-post-development-site sampling when attempting to determine water quality effects from land development projects. The U.S. Environmental Protection Agency, U.S. Geological Survey, and many other agencies can provide guidance on water quality sampling procedures.

Step 3. Identify habitats of threatened and endangered species

The regional or municipal planning body should recognize any threatened or endangered species and their habitats to guarantee their survival. Planners should use this information to choose areas where development can occur without endangering these species further. If these species' preferred habitats are first identified on a regional basis, it is encumbent on the developer not to disturb them by construction. More information on this subject can be obtained from the U.S. Fish and Wildlife Service, the state fish and wildlife department, and local naturalists. In some areas, the state museum can help.

Step 4. Analyze adjacent land uses

Urbanization and other land use effects on aquatic ecosystems have been discussed in Chapter 3 and elsewhere. At the regional level, consider type, extent, and vegetative cover distribution on watersheds, including forests, orchards, and crop areas, the extent to which the region has been developed for residential and industrial purposes, and the presence of undeveloped floodplains and wetlands preservable through sound planning. Open space like wildlife refuges, parks, and golf courses should be identified and plotted on maps. Likewise, land uses obviously causing accelerated erosion or severe pollution should be identified and the areas or water bodies affected should be spotted on maps.

In analyzing land use data, consider housing density, imperviousness extent tolerable by aquatic systems in different areas, possible urbanization effects on ground water recharge and water table, potential fertilizer and pesticide effects from runoff carrying nutrients and pollutants into water bodies, and waterfowl use of a suburban wetland.

Step 5. Identify limiting factors

Limiting factors on fish and wildlife populations include disease, parasites, predation, fishing, hunting, trapping, vehicular accidents, food availability, cover, water, and living space. Water pollution is also a limiting factor, both at regional and municipal levels. Sections 208 and 303E of the 1972 amendments to the Federal Water Pollution Control Act (PL 92-500) provide for water quality studies on a regional scale to identify potential water pollution sources and prescribe methods for stemming these loads. However, unless performed at municipal or township levels, for which an inventory similar to that prescribed for site design planning is made, determining limiting factors on a regional scale is difficult. At the municipal level toxic industrial wastes discharged to receiving streams with resultant fish die-offs or fish rendered non-edible, e.g., by PCBs, may be a limiting factor. Such discharge sources or those of heavy nutrient loadings affecting water quality, fishing, or other recreational opportunities should be identified and measures taken to correct the problem. Dams or other obstructions in streams should be noted. For example, after water quality in Philadelphia's Schuylkill River was improved through pollution control, returning shad were unable to move upstream to spawn because of a dam within the city. Construction of a fish ladder at the dam site permitted fish passage and shad are now spawning upstream.

Step 6. Develop a continuous open space-wildlife corridor system

Considerable attention has been paid to open space corridor development for terrestrial wildlife[180] and to environmental values.[181] There are many opportunities for using rivers, streams, and other water bodies as bases for such open space systems. Once the aquatic resources within a region or municipality have been delineated, consideration should be given to their values in developing them. Vegetated river floodplains offer a ready-made buffer because they are unsuitable for development. Aquatic areas around municipalities may be interconnected to create lakes, for example, as in the Detroit Metroparks previously mentioned, thereby increasing their utility for canoeing or other water-based recreation.

Step 7. Review the development plans

Reviewing development plans avoids adverse impacts on aquatic resources. Because the review comes before development begins, developers are receptive to a municipality's suggested changes and recommendations to expedite building permits.

Adverse effects occurring during and after the construction phase should be considered, including onsite and off-site effects downstream or within the watershed. Natural constraints imposed by the landscape like soil erodibility and steep slopes, and by the aquatic resource, e.g., floral and faunal habitat requirements should be recognized and considered in the review. Data from the water resources and land use inventory can serve as the review's focus. During the review, concentrate on the following:

- Is construction necessary?
- Is the project sited in an appropriate area insofar as natural environmental constraints imposed by the landscape and the aquatic resource are concerned?
- Is the project designed to minimize site and off-site environmental disturbance to water bodies?
- Is erosion or pollution control needed, and will the proposed methods be effective?

A checklist of items, for careful evaluation during development plan review follows. Many points apply to site and municipal plans alike.

(a) Legal requirements and permits (local, state, and federal)

Operations or activities governed by law or requiring permits include:

- dredging and filling
- structural modifications to streams, rivers, and wetlands
- surface water withdrawals
- sediment control
- point source discharges
- land sewage disposal systems
- construction on floodplains
- stream crossings

When in doubt about permits, county or state environmental departments should be consulted. Section 208 plans of the Federal Water Pollution Control Act will be helpful in the review.

(b) Modification to groundwater levels and/or flow patterns

The reviewers should scrutinize proposed water retarding structure placement in relation to surrounding vegetation and consider a development's effect on groundwater levels.

(c) Export of water across watershed boundaries

Surface and groundwater withdrawal from a water body or drainage area can adversely affect the water budget of a watershed when water is not restored, thereby affecting fish and other aquatic life. Regional treatment systems in which wastes are transferred across watershed boundaries can contribute to pollution problems.

[180](162) D. L. Leedy, R. M. Maestro, and T. M. Franklin, *Planning for Wildlife in Cities and Suburbs.*

[181]P. Lewis, *Regional Design for Human Impact,* Thomas Publications, Kaukauna, WI, 1969, 307 pp.

Figure 35. *This is ponded sewage effluent from a septic field on soil with severe limitations for such use. It has a disagreeable odor, and it is a sanitation hazard. When tile lines are punched with a probe, the effluent rises to the surface.*

(d) Erosion and sedimentation

Large development projects should be excluded from steep slopes and where soils are highly erodible. Prompt effective erosion and sediment control measures, as previously discussed, should be used.

(e) Sewage disposal practices

Sewage disposal via septic tanks, spray irrigation systems, package treatment plants, or regional treatment plants depends on various factors and whether soils will assimilate waste. Septic tanks are not suited for moderate to high density housing. Climate, slope, soil, vegetation, and design features should be carefully evaluated before instituting land treatment systems. Back-up systems are necessary during winter in some areas. Individual states usually issue permits for sewage waste disposal and can provide some guidance in evaluating the relative merits of sewage treatment options in an area. Section 201 of the Federal Water Pollution Control Act requires use of the best practical waste water treatment technology in any plant built with federal assistance and mandates the study of alternative waste management techniques before any project is funded.

Sewering practices associated with large treatment systems can cause many short-term effects during sewer construction and can have major long-term ecological ramifications because these lines will dictate future growth patterns and resultant pollutant loading.[182] Sewer lines should be directed away from unique or critical areas of fish and wildlife use. Alternatives for small wastewater treatment systems[183] should be employed in environmentally sensitive areas if development is necessary.

(f) Combined vs. separate sewering

Separate sewers are considered better than combined stormwater-sewage systems which often permit untreated sewage to enter a receiving water body at times of heavy runoff.

(g) Stormwater runoff

All development projects should include a plan specifying ways to regulate runoff release including adequate vegetative buffer strips along waterways.

(h) Nonpoint source pollutants

Evaluate the plan for methods minimizing nonpoint pollution of aquatic areas. Planning under Section 208 of the Water Pollution Control Act is done on a regional basis and deals with nonpoint as well as point sources of pollution.

(i) Construction in floodplains

For safety reasons and to prevent severe stream degradation, construction should be avoided in floodplains (minimum 100-year floodplain). At the same time, the plan

[182](126) T. Dolan, IV and R. M. Maestro, *The Environmental Assessment Statement as a Natural Resource Planning Tool.*

[183]U.S. Environmental Protection Agency, *Alternatives for Small Wastewater Treatment Systems—On Site Disposal-Septic Management and Disposal,* EPA Technology Transfer Seminar Publication, EPA 629/4-7, 1977, 90 pp.

could provide for certain kinds of recreation use of floodplains, e.g., parks and nature trails.

(j) Channel modification including channelization and shore line alteration

Keep to a minimum, except for restoration of previously altered channels.

(k) Wetland drainage, dredging and filling

Wetlands should be preserved wherever possible. Permits are required by some states for construction in wetlands and soggy areas. When in doubt about wetland occurrence, state regulatory authority should be consulted. The Corps of Engineers is the permitting authority on dredge and fill permits.

(l) Potential nuisance problems

The plan should be evaluated from the standpoint of nuisance plant and animal populations, like algae and mosquitoes, which construction, sewage and stormwater management practices may encourage. Native rather than exotic plant species should be used for erosion control.

(m) Alteration to natural flow and circulation patterns

Avoid stream diversion and design properly sized culverts to insure flushing action and fish passage. Docks and piers should not restrict circulation in estuarine or coastal areas. Evaluate dams regarding their need for fish ladders and exercise care in the use of dikes in wetland areas to ensure that they will accomplish water level control without damaging other aquatic resources.

(n) Impoundments

Where impoundments are proposed, consider downstream effects on water quantity and quality.

(o) Buffer strips

Make certain that vegetative buffer strips along water courses are included in the development plan.

(p) Forest clearance

Determine whether forestlands' scheduled clearance will degrade cold water fish habitats. Selective clearance, large lot acreage, or home site clustering offers a more ecologically satisfactory solution to development in such areas.

(q) Scheduling of construction project

When reviewing development plans consider construction project scheduling. Where valuable aquatic environments occur near a development site, schedule construction so that disturbances during fish spawning and waterfowl usage are minimized. When possible, permit con-

Photo: USDA, Soil Conservation Service

Figure 36. *This storm sewer outlet, emptying only 50 or 60 feet in back of a stream, causes ponding and health hazards.*

Figure 37. *Many back yards in the center of this unusually large housing development, built on formerly wooded land, have nearly vertical slopes and severe erosion problems. (Dover, New Jersey)*

struction only when disturbed areas can be quickly revegetated to reduce erosion and water pollution. Recognize cumulative construction effects where a series of projects are in progress. While one development project may have only minor impact on aquatic resources, many simultaneous projects could have devastating effects. A stream's carrying capacity is limited, so avoid unnecessary projects. Approved construction that is properly sited, designed to minimize ecological disturbances, and which adheres to effective sediment and control practices, will use a stream's capacity to maximum advantage. See Appendices G, H, I, and J for information about inventorying wetlands, streams, rivers, lakes, and ponds, together with criteria for various water quality characteristics which may be helpful in the review process.

Step 8. Implement the aquatic resources development plan

There are many land use regulatory tools usable by local, county, and regional governments, and private organizations to implement aquatic resource protection while serving other public interests. Many states have empowered local governments to regulate land use but apparently most local jurisdictions have not exercised these powers. The U.S. Environmental Protection Agency[184] cited one authority [185] who estimated that only 5,000 out of 60,000 jurisdictions having land use powers exercised them in 1974. That adoption of ordinances and use of police powers invested in local ordinances can be effective is attested to by Thurow et al.[186] who give examples for environmentally sensitive areas. Some approaches for land use control include zoning, critical area designation programs, environmental impact statement reviews, performance oriented controls, acquisition of property interests, taxation and charges, and land stewardship programs.

(a) Large lot zoning

Specifying minimum lot-size requirements can serve several water quality purposes. Large lot zoning might be justified for some suburban or rural areas where environmentally significant resources like trout streams exist, but the rationale for such zoning must be carefully evaluated.

(b) Zoning for protection of open spaces

The values of open space areas for recreational and aesthetic amenities and other purposes are well recognized. Courts have usually upheld open space programs instituted for health and safety reasons; therefore, floodway restrictions and buffer zone preservation have a high rate of success under such programs.[187]

(c) Rezoning

Where development has not begun, rezoning might be the simplest way to control situations adversely affecting aquatic resources. However, rezoning and "spot zoning," especially, can be viewed as arbitrary or discriminatory by

[184] U.S. Environmental Protection Agency, *Legal and Institutional Approaches to Water Quality Management Planning and Implementation*, Contract No. 68-01-3564, 1977.

[185] P. Wolf, *The Future of the City: New Directions in Urban Planning*, Watson-Guptill Publications, Whitney Library of Design, New York, 1974.

[186] C. Thurow, W. Toner, and D. Erley, *Performance Controls for Sensitive Lands: A Practical Guide for Local Administrators*, Reports 307 and 308, Planning Advisory Service, 1975. American Society of Planning Officials, 1313 E. 60th Street, Chicago, IL 60637.

[187] (184) U.S. Environmental Protection Agency, *Legal and Institutional Approaches to Water Quality Management Planning and Implementation*.

Figure 38. *A happy angler with two large catfish taken from a reservoir.*

courts when not in conformance with a comprehensive plan or when the basis for change is not clearly identified.[188] [189] From a regional standpoint, rezoning can help preserve or enhance aquatic resources having broad geographic significance. Along coastal areas, it can discourage dumping.

(d) Comprehensive design zones

Various comprehensive design zones (CDZ) can be used to encourage preservation of environmental features. The developer, to achieve maximum densities, must incorporate open space, natural features, recreational opportunities, buffers, and other significant environmental features in his design.[190] The CDZ is particularly effective for planned unit developments (PUDs) and new towns.

(e) Critical area designation programs

These programs preserve ecologically sensitive resources such as floodplains, wetlands, and coastal zones. They have been effective, also, in erosion control, hillside protection, and for protecting waterfowl habitat, fish spawning areas, and habitats of threatened and endangered species.[191] Growth must be carefully controlled or prohibited in such areas. Protection is regulated by the Endangered Species Act administered by the USDI's Fish and Wildlife Service.

(f) Environmental impact statement review

This land use control technique provides a means for evaluating environmental consequences of specific developments. Because the statement is prepared before permit approval and construction, it is effective in identifying special precautionary measures needed for individual projects.[192] [193] It can be used by state, regional, and local agencies which can impose EIS requirements for permits. For example, California requires EISs on private actions licensed by local government.

(g) Performance-oriented controls

These are important when incorporating aquatic resource protection measures into a development program. Because many states endow local government with power to regulate subdivisions, enabling legislation often is not a primary issue. For performance-oriented controls to be effective, however, commitment and involvement by the municipality is necessary. The municipality must be prepared to issue statements regarding environmental goals and provide the developer standards or measurable environmental disturbance levels acceptable in implementing the public policy goal. Performance-oriented and related controls, including local ordinances, have been described.[194] [195] The Brandywine Conservancy, in Chester County, Pennsylvania, and Medford Township, New Jersey have prepared and implemented ordinances for protecting water and related natural resource values.

(h) Acquisition of property interests

These can be used to manage growth and control location of developmental activities. Fee simple acquisition is the most complete type of control but is often more expensive than acquiring partial interests such as that exercised over the development type permitted. Programs can also be aimed at permanent or temporary acquisition. Several, including land banking and the transfer of development rights (TDR), have been used in some jurisdictions to help preserve prime agricultural land while providing tax relief to farmers. Such schemes postpone or eliminate developmental pressure to accommodate more orderly planning, and can help preserve or manage aquatic resources.

Acquisition programs are costly. Bonds approved by public referendum, special taxes like real estate transfers, and some federal programs, such as the water bank program of the U.S. Department of Agriculture or those under the Housing and Community Development Act of 1974, can give some financial relief to municipalities in acquiring property interests. Towns have used this tool for scenic easements, open space programs, prime agricultural land preservation, and critical area protection, including buffer strips along water bodies.

(i) Taxation and charges

Public improvement costs, such as sewage waste treatment plants, can be met by taxation and service charges. Preferential taxation schemes have been used in some states to preserve prime agricultural land, thereby deterring development.

(j) Land stewardship programs

The private sector can control or limit growth in areas having special natural resources or scenic and historic values through stewardship. Preservation and conservation easements are probably the most common means of accomplishing these and related objectives.[196] Private conservancies have helped preserve natural and historic areas, including aquatic resources.

See Chapters 8 and 9 for additional information.

[188](184) U.S. Environmental Protection Agency, *Legal and Institutional Approaches to Water Quality Management Planning and Implementation.*

[189]M. J. Meshenberg, *The Administration of Flexible Zoning Techniques.* pp. 33-42 in Management and Control of Growth—Techniques in Application, Vol. IV, F. Schnidman, J. A. Silverman, and R.C. Young, 1978, 353 pp. (Urban Land Institute, Washington, DC.)

[190]I. J. Parker, *Comprehensive Design Zone: Using Zoning to Protect the Environment,* pp. 48-51 in Management and Control of Growth—Techniques in Application, Vol. IV.

[191]U.S. Department of the Interior, *Methods and Techniques for Critical Area Program Development: Technical Supporting Report A 79,* 1945.

[192](184) U.S. Environmental Protection Agency, *Legal and Institutional Approaches to Water Quality Management Planning and Implementation.*

[193]A. Jokela, *Self Regulation of Environmental Quality: Impact Analyses in Local Government,* Undated. (Center for California Public Affairs, Claremont, CA.)

[194](186) C. Thurow, W. Toner, and D. Erley, *Performance Controls for Sensitive Lands: A Practical Guide for Local Administrators.*

[195]Bucks County Planning Commission, *Performance Zoning,* 1976, 66 pp. (Bucks County Planning Commission, Doylestown, PA 18901.)

[196](119) W. H. Whyte, *The Last Landscape.*

Part IV

Sources of Technical and Financial Assistance and Recommended Readings

Chapter 8

Technical Assistance

Individuals and groups like Trout Unlimited or the Audubon Society are dedicated to improving fish, wildlife, and water-oriented recreation programs, and often lead in implementing improvement plans. Technical assistance sources to which they may refer are identified in this chapter.

Conservation Directories and Other Information Source Books

The National Wildlife Federation's "Conservation Directory" is probably the most comprehensive helpful single document for obtaining names, addresses, and telephone numbers of public and private organizations, agencies, societies, and officials concerned with natural resource use and management. It is updated and published annually and may be purchased from the National Wildlife Federation, 1412 16th Street, N.W., Washington, DC 20036. Each agency and organization listed has a brief description of its functions or purposes. In addition to U.S. federal departments and agencies—including some regional as well as Washington offices—state fish and game commissioners, directors, international, national and interstate commissions, and many private conservation organizations are listed.

The "U.S. Government Manual," available from the U.S. Government Printing Office, Washington, DC 20402, describes functions and services rendered by federal departments, agencies, and offices, and gives their addresses.

The "Catalog of Federal Domestic Assistance," available from the Superintendent of Documents, is updated annually by the President's Office of Management and Budget, and lists all federal financial support available together with a contact.

The "Book of the States," published by the Council of State Governments, Iron Works Pike, Lexington, KY 40505 provides information on state agencies and their responsibility.

The HCRS Information Exchange, Heritage Conservation and Recreation Service, 440 G Street, N.W., Washington, DC 20243, produces "Technical Assistance Notifications" as useful items are collected. It includes materials organized by interest groupings and covers reports, surveys, training manuals, handbooks, and the like that solve specific problems. Document availability is indicated in the descriptions. A recently initiated HCRS service, it should prove valuable to planners.

Technical Assistance According to Selected Categories

Planning for urban fishing and other waterfront recreation requires a broad information base extending from physical, chemical, and biologic environmental components and people's recreation demands, to economic and legal factors influencing the plan. Though the information and technical assistance sources list presented here is not exhaustive, it may prove useful to planners and developers.

1. Fish and other aquatic wildlife

An information search on fish, other aquatic wildlife, and their habitats might well start by contacting the American Fisheries Society, American Society of Ichthyologists and Herpetologists, American Society of Limnology and Oceanography, American Ornithologists' Union, Inc., Ecological Society of America, and The Wildlife Society, all of which are listed in the National Wildlife Federation's "Conservation Directory." They can identify members in a given locality who can offer assistance. Local members of the Izaak Walton League of America, Trout Unlimited, the Audubon Society, and similar organizations, may also supply useful information. The local telephone directory can often furnish a professional society contact. The county agricultural agent or extension service might be helpful, too. State fish and game or conservation departments (see Appendix K) will also have information on fish and wildlife, their requirements, and management.

The Sport Fishing Institute, Suite 801, 608 13th Street, N.W., Washington, DC 20005, has encouraged national development of urban fishing programs. And the Sport Fishery Research Foundation (same address) supports some research in the sport fishery resource field. The Wildlife Management Institute, 1000 Vermont Avenue, N.W., 709 Wire Building, Washington, DC 20005, encourages waterfowl, nongame, and other research-management activities and promotes conferences. Transactions of the annual "North American Wildlife and Natural Resources Conference," often containing articles on aquatic resources and urban wildlife, are available from the Institute.

The Fish and Wildlife Service, U.S. Department of the Interior, Washington, DC 20240, the National Marine Fisheries Service and Office of Coastal Zone Management—components of the National Oceanic and Atmospheric Administration—U.S. Department of Commerce,

Washington, DC 20230, are involved in fish and wildlife conservation programs. The U.S. Fish and Wildlife Service (with the cooperation of the National Marine Fisheries Service), for example, is the agency charged with implementing the Endangered Species Act of 1973. It also operates national refuges and fish hatcheries and develops information on fish and other aquatic wildlife, including methods of rough fish control. Fish stocking and fishery stocks data are available from state conservation departments, the Fish and Wildlife Service, and from the U.S. Department of Agriculture, Soil Conservation Service, Washington, DC 20250. Regional offices for these agencies are listed in the Appendices.

2. Water resources, including pollution and flood control

Many federal departments, agencies, and offices are involved with water resources programs, but only a few are mentioned here. The U.S. Water Resources Council, at 2120 L Street, N.W., Washington, DC 20037, encourages the conservation, development, and utilization of water and related land resources on a coordinated basis by federal, state and local governments and by private enterprise.

There are several water-oriented agencies within the U.S. Department of the Interior. Among these are the Office of Water Research and Technology which, in addition to sponsoring research at state universities and elsewhere, operates a Water Resources Scientific Information Center having access to computerized information on water resources—including recreation values—nationwide. The U.S. Geological Survey, National Center, Reston, Virginia 22092, has information on water quality and on ground and surface water supplies. The Water and Power Resources Service is concerned with reservoir recreation use, irrigation, and power.

Considerable attention has been given to the U.S. Environmental Protection Agency, 401 M Street, S.W., Washington, DC role in water pollution control programs. Regional offices from which information can be obtained about planning for urban fishing and other water-related recreation are listed in Appendix B. Other federal agencies whose programs or services are important to such planning include the U.S. Department of Agriculture and the Corps of Engineers, both of which have major roles in flood control.

Pertinent reports issued or prepared by these agencies include:

(1) From the U.S. Water Resources Council—

(a) "State of the States: Water Resources Planning and Management," dated April 1980, identifies state by state, statutes, policies, and regulations dealing with comprehensive water resources planning and management and summarizes some of the water problems and initiatives being taken within the states to solve them. Names and addresses of useful contacts are provided.

(b) "The Nation's Water Resources 1975-2000: Second National Water Assessment," makes many references to water-related recreation.

(2) From the U.S. Geological Survey, in addition to current data on water quality and quantity—

(a) "Hydrology for Urban Land Planning—A Guidebook on the Hydrological Effects of Urban Land Use."[197] Though somewhat dated, this guidebook will help urban planners.

(3) From the U.S. Environmental Protection Agency (U.S. Government Printing Office)—

(a) "Quality Criteria for Water"[198] describes pH and pesticide effects, etc. on aquatic life.

These agencies and others, including organizations such as river basin commissions listed in the "Conservation Directory," are interested in urban planning in relation to water resources.

State geological surveys, state engineers, environmental protection agencies and other agencies can provide valuable information for planning. Likewise, private organizations like the Conservation Foundation, 1717 Massachusetts Avenue, N.W., Washington, DC 20036 and Partners for Livable Places, 2120 P Street, N.W., Washington, DC are interested in urban aquatic resources planning. The latter has compiled a directory of participants in an urban waterfront action group (see Appendix L).

3. Soils, minerals, and soil erosion control

The U.S. Department of Agriculture, through its Agricultural Research Service, Washington, DC 20250, researches water and soil management. The Soil Conservation Service (same address) dispenses resource data and technical assistance for soil and water conservation projects in both rural and urbanizing areas. Most of the nation's 3,000 conservation districts, formed as legal state government subdivisions, cooperated with HCRS and SCS in inventorying privately owned recreation facilities in each district. They guide local citizens in the use of government programs and services to increase recreation opportunities, and encourage participation by local individuals, groups, and organizations throughout program planning. Information about district operations and projects sponsored are given in "Working Together in Recreation."[199]

Soil testing is performed by the U.S. Agricultural Extension Service office in each state and by state university and private laboratories. The U.S. Geological Survey contributes information on mineral resources and offers many different map types useful in planning. The National Sand and Gravel Association, 900 Spring Street, Silver Spring, MD 20910, in publications dealing with progressive rehabilitation of areas from which sand and gravel have been extracted, suggests ways for using the areas for fish, wildlife, and recreation. Information on the latter use is available, also, from the Heritage Conservation and Recreation Service.

4. Vegetation and its management

The U.S. Department of Agriculture's Agricultural Research, Forest, and Soil Conservation Services have

[197](165) L. B. Leopold, *Hydrology for Urban Land Planning—A Guidebook on the Hydrological Effects of Urban Land Use.*

[198]U.S. Environmental Protection Agency, *Quality Criteria for Water,* G.P.O. No. QCW-055-011, 01049-4, 1976, 256 pp. (U.S. Environmental Protection Agency, Office of Water and Hazardous Materials, Washington, DC.)

[199]National Association of Conservation Districts, USDI Heritage Conservation and Recreation Service, and USDA Soil Conservation Service, *Working Together in Recreation,* Undated, 18 pp. (NACD Service Dept., 408 E. Main, P.O. Box 855, League City, TX 77573).

publications on preserving valuable trees during construction, stabilizing stream banks, and vegetation's role in controlling erosion.

The U.S. Fish and Wildlife Service has had extensive experience in planning and managing wetland areas in connection with its system of National Wildlife Refuges. Much of the management is through waterlevel manipulation. This Service, the Water and Power Resources Service, the Army Corps of Engineers and other agencies have dealt with nuisance aquatic plant control problems. The National Audubon Society, 950 Third Avenue, New York, NY 10022 manages many sanctuaries in aquatic areas and, like the agencies mentioned above, can render valuable advice.

As an aid in making environmental assessments and in inventorying aquatic habitats, a recent Fish and Wildlife Service publication, "Classification of Wetlands and Deepwater Habitats of the United States"[200] is suggested. The document includes definitions, descriptions, keys, and illustrations encompassing both freshwater and estuarine areas.

5. Socio-economic, legal, and institutional aspects

Socio-economic, legal, and institutional aspects of water resources planning and management can be identified and obtained as abstracts from the Water Resources Scientific Information Exchange, Office of Water Research and Technology, U.S. Department of the Interior, Washington, DC 20240. These are based on research by federal and state agencies, universities, and private organizations. Among private nonprofit organizations active in this area is Resources for the Future, Inc., 1755 Massachusetts Avenue, N.W., Washington, DC 20036.

Particularly interesting to urban and regional planners concerned with wetlands is "Strengthening State Wetland Regulations."[201] This report, prepared for the Fish and Wildlife Service by the Environmental Law Institute (1346 Connecticut Avenue, N.W., Washington, DC 20036) assists states in developing and strengthening wetland protection programs. It deals in part with the Clean Water Act of 1977 relating to discharge control of dredged and fill materials into waters, including wetlands. An overview of state programs is provided and appendices contain valuable information on floodplain and wetland regulations.

Information on past, current, and predicted use of aquatic resources for fishing and other forms of outdoor recreation is available from the U.S. Department of the Interior, the U.S. Water Resources Council, and the Sport Fishing Institute. The Fish and Wildlife Service, the Heritage Conservation and Recreation Service, the Forest Service, and many states and universities are giving increased attention to people's preferences and needs for water-based recreation.

Flooding is economically significant in urban development. Details on the federally subsidized National Flood Insurance Program are available from the Federal Insurance Administration, U.S. Department of Housing and Urban Development, Washington, DC 20410.

Other federal laws relating to communities, planners and developers will be considered in the next chapter and relate to authorizations for financial assistance.

Universities, private research organizations, and consultants are also good information sources.

The Role of Consultants

Few planning firms and development corporations have all the expertise needed to make aquatic resource inventories and develop "from scratch" all other information required to formulate and implement a sound aquatic resource plan. For the individual planner or developer, having detailed knowledge about the physical, chemical, and biological make-up of the aquatic environment, disturbance factors, future projections, performance standard development, applicable ordinances, etc. would be impossible without assistance. When it is not economically feasible to employ full-time staff members to obtain this information, the services of consultants is a good alternative.

Experts in biology, ecology, geology, limnology, hydrology, environmental planning, economics, sociology, engineering, law and other disciplines are available as consultants. Planners must locate the proper person or consulting firm for the task and define the consultant's work explicitly. Obviously, consultants should be qualified to do the work they are employed to do. Many professional societies certify certain members as biologists or whatever. A part of the responsibility for engaging qualified consultants to do a required job, however, rests with the planner or developer, who should look into the background and record of prospective consultants, reviewing progress as work proceeds.

[200](128) L. M. Cowardin, V. Carter, F. C. Golet, and E. T. LaRoe, *Classification of Wetlands and Deepwater Habitats of the United States.*

[201]J. A. Kusler, *Strengthening State Wetland Regulations,* Performed by the Environmental Law Institute for the Office of Biological Services, Fish and Wildlife Service, USDI, Washington, DC 20240, FWS/OBS-78/98, 1978, 147 pp. (U.S. Government Printing Office, Washington, DC 20402).

Chapter 9

Financial Assistance

Funding for urban fishing, water-related outdoor recreation, and waterfront revitalization comes from various public and private sources. Many projects are supported cooperatively by funds from federal, state, and local agencies, private organizations, and industry. Sometimes, donations of land, water, equipment, and personnel are made in lieu of funds. Recognizing that most communities and planners know about local and state funding possibilities, this chapter focuses primarily on opportunities for federal financial assistance, but gives some examples of other funding sources.

Federal Funding Sources

Though concerned federal agencies may be contacted directly for information on funding (see directories listed at the beginning of Chapter 8) there are several available publications presenting overviews of many relevant federal programs. For example, Harney[202] and a more recently published guide to urban waterfront improvement (*Improving Your Waterfront: A Practical Guide,* 1980, Department of Commerce, Office of Coastal Zone Management) review the assistance obtainable from federal agencies for waterfront revitalization and use, including access area acquisition and redevelopment. Grants to the states encourage redevelopment of underutilized waterfronts for recreation, open space, and economic development. Harney and the Commerce Department Guide provide valuable guidance for obtaining assistance from other agencies. Thus, program name, funding categories, primary program focus, types of assistance, qualified applicants, contact and address, legal authorization, and financial information is furnished for:

- Department of the Army, Office of Chief of Engineers
- Commerce Department—Economic Development Administration, Maritime Division, and National Oceanic and Atmospheric Division
- Community Services Administration, Environmental Protection Agency
- General Services Administration
- Department of Housing and Urban Development
- Interior Department—Bureau of Land Management, Heritage Conservation and Recreation Service
- National Foundation on the Arts and Humanities—National Endowment for the Arts
- Small Business Administration
- Department of Transportation—Federal Highway Administration.

Several of these departments have more than one program in this area of interest.

The National Urban Recreation Study[203] identifies several relevant funding sources including:

(1) The Land and Water Conservation Fund is administered by HCRS and is the major categorical grant program for recreation. It provides matching grants to state and local governments for acquisition and development of public parks and recreation areas. These funds can be made available for urban waterfronts, access, and acquisition of stream banks, development of recreational water areas, etc.

(2) The Comprehensive Employment and Training Act has enabled many park and recreation agencies in economically hard-pressed cities to maintain recreation services that would otherwise have terminated if dependent on local funds alone.

(3) The 701 planning program of the 1974 Housing Act, as amended, facilitates urban planning and requires all grant recipients to develop a comprehensive planning process incorporating at least a land use and a housing element to be eligible for continued funding. Under these grants, many communities have prepared and adopted plans for critical areas such as floodplains and wetlands as part of broader planning efforts.

(4) The National Flood Insurance Program offers federally-subsidized flood insurance (up to 90% in participating communities imposing required land use controls on new development). The program discourages new developments in riverine floodways, but allows floodproof construction on the floodplain fringe.

(5) The 1974 Safe Drinking Water Act requires filtration of all surface water supplies presently stored in reservoirs. This could have significant impact on the future of watershed lands. Cities may feel compelled to sell watershed

[202](88) A. L. Harney (ed.), *Reviving the Urban Waterfront.*

[203]USDI Heritage Conservation and Recreation Service and National Park Service, *National Urban Recreation Study,* 1978. (Government Printing Office, Washington, DC 20402.)

lands to defray constructing and operating costs for required filtration systems. Cities may justify water supply land retention in open space only if the land also serves additional public purposes such as outdoor recreation.

(6) The 1965 Federal Water Project Recreation Act mandates outdoor recreation, fish, and wildlife enhancement consideration in planning and developing all federal navigation, flood control, reclamation and multipurpose reservoir projects. It requires state and local coordination in project planning and management, and encourages non-federal management of project lands and waters for recreation and for fish and wildlife purposes. The Act provides up to 50 percent of development costs for state and local recreation facilities on project lands, when non-federal sponsors agree to operate and maintain these facilities. Up to 75 percent of fish and wildlife enhancement costs are authorized. Loans to state and local governments to help match the federal share of development costs are empowered.

(7) The 1977 Housing and Community Development Act authorized a new Urban Development Action Grants program aimed at neighborhood and community rehabilitation in severely distressed cities. Parks and recreation projects could play a significant role in this urban revitalization program.

(8) Abandoned transportation or utility corridor conversion to public trail uses provides valuable recreation opportunities while providing options for further use. Many existing railroad rights-of-way, for example, were acquired through direct or indirect public assistance. Yet, when rail service is discontinued, these valuable linear resources are threatened by piecemeal sale for other uses. P.L. 94-210, the Railroad Reorganization and Regulatory Reform Act, Section 809(b) authorizes funds for "rails to trails" conversion funding and provides grants up to 90 percent of the planning, acquisition, and development cost for abandoned railroad rights-of-way for recreation and conservation.

Fish hatcheries are attractive as a recreational pursuit but have generally been located in rural areas. Since the inception of the Land and Water Conservation Fund, HCRS has funded 30 fish hatcheries in 16 states. The federal grant for these projects totals $16.2 million. HCRS recommends locating fish hatcheries near urban areas to facilitate visits by urban citizens. Priority consideration for funding will be given to projects meeting this criterion.

In 1977, Michigan located a hatchery near Detroit for stocking the Detroit River and Lake St. Clair. HCRS provided $3.5 million for this project.

Section 73 of the 1974 Water Resources Development Act (P.L. 93-215) authorizes 80-20 funding for acquisition of floodplains for recreation, fish and wildlife, and other public purposes. To date, however, it has provided limited support for waterfront projects.

The Office of Coastal Zone Management (OCZM) has developed programs offering planning grants for waterfront development. State coastal zone grants can be used for comprehensive waterfront planning to support planning, design, and engineering studies for urban waterfront revitalization, including protection and restoration of historic, cultural, and aesthetic resources. Industrial and port development and increased coastal access for public recreation purposes is included.

The statute reauthorizing the Coastal Zone Management program passed by Congress October 1, 1980, contains new provisions in Section 306A authorizing grants to eligible states having approved coastal zone programs for redevelopment of deteriorating and underutilized urban waterfronts and ports, provision of access to public coastal areas, and the preservation and restoration of certain other particularly valuable coastal areas. The grants, authorized up to $20,000,000 per year, may be used for purchasing land, low-cost construction projects, pier rehabilitation or acquisition, shoreline stabilization, piling replacement or removal, certain studies, and other costs. No appropriation has yet been made under the section, but regulations will be developed to implement provisions. Contact the Urban Waterfront Program (202) 634-4245.

The Office of Coastal Zone Management/Coastal Energy Impact Program (OCZM/CEIP) awards energy impact grants and loans to eligible coastal states to assist local communities to plan for and deal with impacts from coastal energy development. The program provides: planning grants related to all types of energy development; construction grants for public facilities supporting population increases stemming from Outer Continental Shelf (OCS) energy development; environmental and recreational grants required to mitigate OCS energy development, oil, gas, coal transportation, transfer, and storage impacts; and alternative ocean energy activity impacts. These funds are administered by state agencies selected by the Governors. Contact Director, Coastal Energy Impact Program (202) 254-8000.

The OCZM also awards grants to coastal states to plan and manage fish stocks and develop improved facilities for fish processing. Contact Director, Coastal Zone Management Programs (202) 634-1672.

The National Estuarine Sanctuary Program was established through the 1972 Coastal Zone Management Act as amended in 1976. This program makes 50 percent matching grants to coastal states for acquiring, developing, or operating estuarine areas set aside "to serve as natural field laboratories in which to study and gather data on the natural and human processes occurring within the estuaries of the coastal zone." By 1979 there were seven sanctuaries in operation and several planned. The one known as Old Woman Creek, Ohio, is one of the few comparatively natural estuaries remaining on the heavily populated shores of Lake Erie.[204] Other similar opportunities may exist for such sanctuaries.

The National Endowment for the Arts fosters growth and development of the arts in the United States, preserves and enriches the nation's cultural resources, and provides opportunities for wider experience in all arts. Small seed grants made by the Design Arts Program of the National Endowment for the Arts have demonstrated the potential extant in urban waterfronts for recreation and contributions to the overall livability of our cities.[205]

The publication, "Recreation and Land Use: the Public Benefits of Clean Waters,"[206] describes EPA and joint EPA-

[204]U.S. Department of Commerce and National Oceanic and Atmospheric Administration, *National Estuarine Sanctuary Program*, U.S. Dept. of Commerce (NOAA), 1979, 3 pp.

[205](88) A. L. Harney (ed.), *Reviving the Urban Waterfront.*

[206](107) J. Gerba and B. Hague, *Recreation and Land Use: the Public Benefits of Clean Waters.*

HCRS programs and funding possibilities for bettering urban recreation, particularly in connection with the Clean Water Act. It also discusses methods for enlisting help from other federal, state, and local sources. Planners and community leaders will find this publication to be a very helpful guide. The final rule for the clean lakes program was published in the *Federal Register*, February 5, 1980, volume 45, no. 25, page 7788, which contains policies and procedures governing the provision of federal financial assistance to states for protection and restoration of publicly owned freshwater lakes as authorized by the Clean Water Act (33 U.S.C. 1251 *et seq.* Section 314).

Planners and developers interested in fund-matching opportunities under the Federal Aid for Fish and Wildlife Restoration programs administered by the U.S. Fish and Wildlife Service in cooperation with the states are advised to contact respective state fish and wildlife agencies or Regional Offices of the U.S. Fish and Wildlife Service (see Appendices K and C).

The Water and Power Resources Service program of the Department of the Interior is unique in several respects. First, it is limited by Reclamation Law to the 17 contiguous western states and Hawaii. Second, the Water and Power program has been based on the principle of repayment by direct beneficiaries. However, most flood control, recreation, and fish and wildlife enhancement costs are nonreimbursable; they are regarded by Congress as widespread benefits accruing to the general public.

In addition to its regular program for planning, constructing and operating multipurpose water developments, Water and Power administers a loan and grant program offering federal assistance to organizations wishing to construct or improve water resources.

The programs were made possible by enactment of Public Law 84-984 (Small Reclamation Projects Act) and Public Law 84-130 (Distribution System Loans Act). The former encourages state and local participation in water resource project development in 17 western states and Hawaii. The latter enables non-federal construction of distribution systems on federal projects, in lieu of construction by Water and Power.

Projects planned and constructed under these loan programs are similar to, but generally smaller than, those under the regular program.

For both programs, the loan applicant designs, builds, operates, and maintains the project. Water and Power's role assures the Secretary of the Interior that projects are feasible, from a financial engineering, and environmental point of view, and represent a reasonable risk for loans under tha acts.

Grants are also available under Public Law 84-984 for those project portions that are nonreimbursable as a matter of national policy. Grants may be made for flood control, recreation, and fish and wildlife improvement where these benefit the general public. Grants may be made even if no loans are requested, provided the project is multipurpose, and that the applying organization can demonstrate its ability to contribute to project cost.

Those interested in opportunities under the Water and Power Resources Service's loan and grant program should contact any Water and Power office. More detailed information regarding loan applications and accompanying documents is contained in brochures entitled "Guidelines for Preparing Applications for Loans and Grants Under the Small Reclamation Projects Act—P.L. 84-984" and "Guidelines for Approval of Applications or Funding of Loans Under the Distribution Systems Loan Act." Copies are available to prospective applicants.

Section 314 of the Clean Water Act (P.L. 95-217) directed the United States Environmental Protection Agency to assist states to control pollution sources adversely affecting the quality of freshwater lakes and in restoring their quality. Through the Clean Lakes Program, EPA provides technical and financial assistance to the states to: (1) classify publicly owned freshwater lakes according to trophic condition; (2) conduct diagnostic studies of specific publicly owned lakes and develop feasible pollution control and restoration programs for them; and (3) implement lake restoration and pollution control projects. To date, over $60 million has been awarded for 275 projects in 45 states. More information may be obtained by calling EPA's Clean Lakes Program (202) 472-3400.

Sources of State and Private Financial Assistance

There are many cooperative programs in which states and local communities can receive financial assistance from the federal government. Yet, the initiative and much of the support for urban fishing and other waterfront recreation programs are from state and local levels. In this guide, bond issues, earmarking of sales tax funds, and use of state-appropriated funds have been mentioned, as have tax incentives, easements, and transfer of development rights as means of conserving or acquiring land and water areas. Chase[207] suggests "One of the most effective nongovernmental techniques is the establishment of land trusts held by community-formed nonprofit corporations. These trusts can receive charitable gifts of land that offer tax benefits to donors and establish community-owned property, perhaps for a garden, a park, or a basketball court. A national organization, the Trust for Public Land (TPL) helps neighborhood groups establish land trusts and develop community-oriented projects on their properties. . ."

Another private organization, the Nature Conservancy, Suite 800, 1800 North Kent Street, Arlington, VA 22209, raises funds to protect ecologically unique or unusual land, e.g., habitats for rare and endangered species or remnant biotic communities.

Other possible funding sources include preservation groups like the Izaak Walton League of America Endowment, Inc., P.O. Box 535, LaGrange, IL 60525, private industry, philanthropic corporations or foundations such as Ford and Rockefeller, and conservation action groups like the National Audubon Society and the National Parks and Conservation Association, 1701 18th Street, N.W., Washington, DC 20009.

[207]J. T. Chase, *Recreation for Urban America*, 1979, 38 pp. (National Committee for Urban Recreation, 1302 18th Street, N.W., Suite 301, Washington, DC 20036.)

Chapter 10

Recommended Readings

This chapter includes a few recommended documents supplementing the information and references included in this guide. Abstracts or annotations should enable readers to determine whether first-hand examination of the publications would be helpful.

Bolton, C. 1979. *Citizen's Action Manual: A Guide to Recycling Vacant Property in Your Neighborhood.* Prepared for U.S. Department of the Interior, Heritage Conservation and Recreation Service. 32 pages. For sale by the Superintendent of Documents, U.S. Government Printing Office, Washington, DC 20402.

This describes the National Urban Land Program of the Trust for Public Land (TPL), focusing on how, with encouragement and minimal government assistance, neighborhood groups can turn unused buildings and land into useful assets. Seven recommended steps are: (1) get organized for action; (2)identify lots to be acquired; (3) acquire the land; (4) organize and incorporate your neighborhood land trust; (5) plan and design the site together; (6) prepare and develop the site; and (7) maintain and preserve the community-owned property.

Valuable advice is given on ways to organize a neighborhood land trust, responsibilities of a land trust, means of acquiring land, information on taxes, planning, developing, liability insurance, etc. An important section deals with "how to find the help you need to make your project work."

Carroll, A. nd. *Developer's Handbook.* State of Connecticut Department of Environmental Protection, State Office Building, 165 Capitol Avenue, Hartford, CT 06115. 60 pp.

This handbook, financed by NOAA under the Coastal Zone Management Act of 1972, while focusing on design and construction considerations for Delaware, has much to offer planners and developers as a general guide to subdivision developments elsewhere. Major natural systems and resources are briefly explained, with opportunities and limitations they impose on development summarized. Particular attention is given to regulatory programs of the State's Department of Environmental Protection, but users are advised to contact local governments and other state agencies to determine other legal requirements. A system for evaluating characteristics of potential development sites is outlined. The handbook includes several pages of sketches with accompanying paragraphs explaining subdivision design and construction practices. Left-hand drawings represent improper methods of development that often result in environmental problems; those to the right illustrate recommended development practices. Sources of further assistance are provided.

Clark, J. 1974. *Coastal Ecosystems: Ecological Considerations for Management of the Coastal Zone.* The Conservation Foundation, 1717 Massachusetts Ave., N.W., Washington, DC 20036 (in cooperation with the National Oceanic and Atmospheric Administration, Office of Coastal Environment, U.S. Department of Commerce), 178 pp.

"The purpose of this guidebook is to reduce a vast stockpile of ecological data to a few simple principles, by means of which to improve our use of coastal lands and waters." The author states that environmental coastal water and shoreland management has as one of its fundamental goals the maintenance of coastal ecosystems as near to the natural condition as possible and that the management program must embrace whole ecosystems, i.e., the adjacent shorelands must be included to the extent they have significant influence on coastal waters. After treating ecological considerations and principles he discusses environmental disturbances, resource evaluation and protection, and constraints on specific uses. Planners and developers will learn much about estuarine areas and factors to consider in any development if such areas are to function naturally.

Fisheries and Environment Canada. 1978. *Guidelines for Land Development and Protection of the Aquatic Environment.* Dept. of Fisheries and Oceans, Fisheries and Marine Service, 1090 West Pender Street, Vancouver, British Columbia, Canada V6E 2PL. Technical Report No. 807. 55 pp.

This publication consists of guidelines identifying problems associated with land development and presents solutions preventing potential deleterious development effects on freshwater river and stream environments. Guidelines are based on Fisheries and Marine Service technical staff experience and knowledge and from a review of pertinent literature. They are divided into three main categories, namely, green strips or buffers along each side of a watercourse, water quantity and quality, and general construction guidelines. Designs for green strips under different situations are illustrated, as are various settlement and detention basins for sediment removal in construction areas, permeable parking lot and roadway medians to promote ground water disposal, and other devices. A copy of the British Columbia Gravel Removal Order governing removal or displacement of gravel in certain fish spawning rivers is included.

Fox, T. 1979. *Land Conservation and Preservation Techniques.* U.S. Department of the Interior, Heritage Conservation and Recreation Service. 75 pp.

This handbook is one of a series prepared to assist government agencies and nonprofit organizations in stretching limited dollars for maximum effectiveness and public benefit. Eleven chapters deal respectively with tax incentives for giving, alternative ways of acquiring property by gift and purchase, detailed examples of alternatives, rehabilitating historic buildings, dealing with corporations, donations and the land and water conservation fund, private land holding organizations, real estate financing, negotiating with the landowner, recreation set-aside legislation, and case studies.

Lackey, R. T. 1975. *Recreational Fisheries Management and Ecosystem Modeling.* Division of Forestry and Wildlife Resources, Virginia Polytechnic Institute and State University, Blacksburg, VA 24061. 44 pp.

Author's abstract: "Fisheries management is the practice of analyzing, making, and implementing decisions to maintain or alter the structure, dynamics, and interactions of habitat, aquatic biota, and man to achieve human goals and objectives through the aquatic resource. The purpose of this article is to place ecosystem modeling into a fisheries management framework, specifically as appropriate to recreational fisheries management. Recreational fisheries are especially complex, but prediction is the essence of fisheries management. Managers usually predict the consequence of a proposed decision in a number of ways including rules of thumb, past experience, population models, experimentation, trial and error, and pure guess. A key problem in making accurate predictions of the consequences of a proposed management decision is the complexity of most fisheries. Arithmetical calculation has been the major problem with using mathematical models in fisheries management. This problem has been solved to some degree by 'simulating' fisheries.

"Most fisheries and ecosystem models are quite similar in approach and philosophy, but there is substantial variation between models when viewed according to their intended use or function. Models used in fisheries may be classified as to habitat, biological, or social type, or combinations of the three categories. Fisheries, when viewed in the broadest sense, includes habitat, biological, and social aspects. The future role of modeling in recreational fisheries management may or may not be great and depends in large measure on the relationship between 'modelers' and 'decision-makers'."

Meshenberg, M. J. 1976. *Environmental Planning: A Guide to Information Sources.* Volume 3 in the Man and Environment Information Guide Series of the American Society of Planning Officials. Gale Research Company, Book Tower, Detroit, MI 48226. 492 pp.

This bibliography on environmental planning contains 1270 references, most of which are briefly annotated. References are numbered consecutively but are arranged alphabetically by author within each of 19 chapters. Chapter titles, indicating the nature of the subjects treated, are: *The Environment and Environmental Issues, Environmental Planning, Environmental Impact Analysis, Environmental Plans, Soil, Geology, Natural Hazards, Water in the Planning Process, Climatology, Vegetation, Wildlife, Population and Urban Growth, Environmental Design, Historic Preservation, Environmental Health, Noise in the Environment, Environmental Law, Energy,* and *The "New" Land-use Planning.* Three indexes are included, by author, title, and subject. Most references occur in the chapter on water in the planning process.

Thurow, C. W. Toner and D. Erley. 1975. (Third Printing, 1977). *Performance Controls for Sensitive Lands: a Practical Guide for Local Administrators.* American Planning Association (formerly American Society of Planning Officials), 1313 East Sixtieth Street, Chicago, IL 60637. Planning Advisory Service Report nos. 307, 308. 156 pp.

This manual identifies key natural processes of five environmentally sensitive areas yielding important public benefits and suggests maintenance techniques using basic police and zoning powers of local government. Typical sensitive areas include streams, creeks, aquifers, wetlands, woodlands, and hillsides. Communities that have adopted ordinances or regulations for such control are listed. These ordinances, important in local land-use control, are well within municipal and county government capabilities. Factors leading to good ordinances are analyzing natural elements associated with environmentally sensitive areas, identifying and mapping the area to be protected, and developing necessary legal language for writing the ordinance. The goal of environmentally oriented land-use regulations is to maintain or preserve natural processes as land undergoes change for man's use. Erosion and runoff ordinances are described, with excerpts of other ordinances for sensitive lands included. The final chapter lists technical assistance resources, many of which were taken from the National Wildlife Federation's "Conservation Directory."

U.S. Army Corps of Engineers, Florida Dept. of Environmental Regulation, and Florida Dept. of Natural Resources. 1977. *State of Florida Joint Permit Application for Dredge, Fill Structures,* 23 pp. + 7 appendices.

Conditions requiring permits are described for construction activities in the waters of Florida. Instructions for completing a single joint application to streamline the permitting process are included. Situations requiring permits are illustrated in color and a flow chart shows processing stages and time passage for final action.

U.S. Department of the Interior, Fish and Wildlife Service, and U.S. Department of Commerce, National Marine Fisheries Service. 1975. *People and the Sound: Fish and Wildlife.* New England River Basins Commission, 270

Orange Street, New Haven, CT 06511. 56 pp. + appendices.

This report, produced in connection with the Long Island Sound Regional Study by the federal government, New York, and Connecticut, is based on studies led by the New England River Basins Commission. It discusses the Long Island Sound (LIS) Region's ecosystem, its fishery, and its wildlife, and recommends measures for using these resources that are compatible on an environmental, economic, and social basis.

The fish and wildlife management plan in this publication (1) avoids conditions or actions leading to reduction in variety or diversity of the ecosystem, (2) discourages actions likely to lead to large-scale irreversible ecological change, (3) places a high value on environmental diversity, and (4) considers ecosystem organization concepts like capacity, succession, specialization, and evolution as integral parts of planning efforts for preserving, protecting, and managing the Long Island Sound region's fish and wildlife resources.

Wetland loss rate in New York and Connecticut appears to be stabilizing near zero as a result of recently enacted legislation there, but water quality problems still remain.

Witherspoon, R. E., J. P. Abbett, and R. M. Gladstone. 1976. *Mixed-use Developments: New Ways of Land Use.* ULI-the Urban Land Institute, Washington, DC 193 pp.

The authors discuss mixed-use development characteristics and advantages where mutually supporting activities are integrated into a single development project. Regional shopping centers grew in size and concept during the 1950's with developers expanding the idea to projects where people could live, walk, shop, play, and visit for entertainment.

"Diversity in mixed-use development means integrating three or more significant revenue-producing uses, for example, office, residential, retail, recreational or hotel (motel) in a pre-planned manner. Each serves as an anchor by individually attracting people for a variety of primary purposes. Combined in a compact physical configuration this mix makes possible the multiple-purpose trip, and substitution of pedestrian for vehicular travel."

Many mixed developments are oriented to waterfront areas. The Watergate development in Washington, DC, had a dual orientation challenge, how to buffer the complex from surrounding blight then characteristic of the area, and how to incorporate the adjacent Potomac River and its beautifully landscaped esplanades into the project. The solution established a "garden-city-within-a-city"—a self-contained environment for living, working, shopping, playing, and cultural functions tied together by a central landscaped plaza facing the river. Building design was essentially curvilinear, in keeping with angles of the surrounding bridges, riverbank, and adjacent Kennedy Center. Curved buildings at the Watergate were arranged to provide maximum exposure to the Potomac.

Development of Marina City, a mixed-use project on the Chicago River's north bank in the heart of downtown Chicago, made the inclusion of a 700-berth marina possible.

Appendices

Appendix A

State and Federal Coastal Zone Management Program Offices

STATE OFFICES

North Atlantic Region

Connecticut

Director, Coastal Area Management Program
Department of Environmental Protection
71 Capitol Avenue
Hartford, CT 06115
203/566-7404

***Maine**

State Planning Office
Resource Planning Division
189 State Street
Augusta, ME 04333
207/289-3155

***Massachusetts**

Program Manager
Executive Office of Environmental Affairs
100 Cambridge Street
Boston, MA 02202
617/727-9530

New Hampshire

Office of State Planning
2½ Beacon Street
Concord, NH 03301
603/271-2155

***New Jersey**

Bureau of Coastal Planning and Development
Department of Environmental Protection
P.O. Box 1889
Trenton, NJ 08625
609/292-9762

New York

Coastal Management Unit
Department of State
162 Washington Street
Albany, NY 12231
518/474-8834

***Rhode Island**

Coastal Resources Management Program
Washington County Government Center
Tower Hill Road
South Kingstown, RI 02879
401/789-3048

South Atlantic Region

***Delaware**

Coastal Management Program
Office of Management, Budget and Planning
James Townsend Building
Dover, DE 19901
302/736-4271

Georgia

Coastal Resources Division
Department of Natural Resources
1200 Glynn Avenue
Brunswick, GA 31520
912/264-4771

***Maryland**

Department of Natural Resources
Tidewater Administration
Tawes State Office Building
Annapolis, MD 21401
301/269-2784

*Indicates federally-approved state programs.

***North Carolina**

Department of Natural Resources
and Community Development
Box 27687
Raleigh, NC 27611
919/733-2293

***South Carolina**

Wildlife and Marine Resources Department
1116 Bankers Trust Tower
Columbia, SC 29201
803/758-8442

Virginia

Council on the Environment
Ninth Street Office Building, Ninth Floor
Richmond, VA 23219
804/786-4500

Gulf/Islands Region

***Alabama**

Executive Director
Coastal Area Board
General Delivery
Daphne, AL 36526
205/626-1880

Florida

Office of Coastal Zone Management
Department of Environmental Regulation
Twin Towers Office Building
2600 Blair Stone Road
Tallahassee, FL 32301
904/488-8614

Louisiana

Coastal Resources Program
Department of Transportation and Development
P.O. Box 44245, Capitol Station
Baton Rouge, LA 70814
504/342-7898

Mississippi

Mississippi Bureau of Marine Resources
Department of Wildlife Conservation
P.O. Box Drawer 959
Long Beach, MS 39560
601/864-4602

***Puerto Rico**

Special Assistant
Department of Natural Resources
P.O. Box 5887
Puerto De Tierra, PR 00906
809/725-2769

Texas

TENRAC
Natural Resources Division
Suite 501
411 West 13th Street
Austin, TX 78711
512/475-0773

***Virgin Islands**

Commissioner
Virgin Islands Dept. of Conservation and Cultural Affairs
P.O. Box 4340
Charlotte Amalie
St. Thomas, VI 00801
809/744-3320

Great Lakes Region

Illinois

Illinois Coastal Zone Management Program
300 N. State Street, Room 1010
Chicago, IL 60610
312/793-3126

Indiana

State Planning Services Agency
143 West Market Street, Harrison Building
Indianapolis, IN 46204
317/232-1482

***Michigan**

Coastal Zone Management Program
Department of Natural Resources
Division of Land Use Programs
Stevens T. Mason Building
Lansing, MI 48926
517/373-1950

Minnesota

State Planning Agency
Capitol Square Building
550 Cedar Street, Room 100
St. Paul, MN 55155
612/296-2633

Ohio

Department of Natural Resources
Division of Water
1930 Belcher Drive, Fountain Square
Columbus, OH 43224
614/466-6557

Pennsylvania

Department of Environmental Resources
Third and Reily Streets
P.O. Box 1467
Harrisburg, PA 17120
717/783-9500

*Indicates federally-approved state programs.

*Wisconsin

Office of Coastal Management
Department of Administration
General Executive Facility 2
101 South Webster Street
Madison, WI 53702
608/266-3687

Pacific Region

*Alaska

Policy Development and Planning Division
Office of the Governor
Pouch AP
Juneau, AK 99801
907/465-3541 (via Seattle Op. 8-399-0150)

*California

California Coastal Commission
631 Howard Street, Fourth Floor
San Francisco, CA 94105
415/543-8555

*Guam

Bureau of Planning
Government of Guam
P.O. Box 2950
Agana, GU 96910
477-9502 (via Overseas Operator)

*Hawaii

Department of Planning and Economic Development
P.O. Box 2359
Honolulu, HI 96804
808/548-4609 (via S.F. Op. 8-556-0220)

Northern Marianas

Office of Planning and Budget Affairs
Executive Office of the Governor
Saipan, CM 96950
6621 (via Overseas Operator)

*Oregon

Land Conservation and Development Commission
1175 Court Street, N.E.
Salem, OR 97310
503/378-4097

*Washington

Department of Ecology
PV-11
State of Washington
Olympia, WA 98504
206/753-4348

California

Bay Conservation and Development Commission
30 Van Ness Avenue, Room 2011
San Francisco, CA 94102
415/557-3686

American Samoa

Development Planning Office
Government of American Samoa
Pago Pago, AS 96799
633-5155 (via Overseas Operator)

Also send state mailings to:

Great Lakes Basin Commission
P.O. Box 999
Ann Arbor, MI 48106
313/668-2320

Coastal Liaison
NE/NY Coastal Task Force
NERBC
53 State Street
Boston, MA 02109
617/223-6244

FEDERAL OFFICES

Regional Manager
3300 Whitehaven Pkwy, N.W.
Washington, DC 20235

North Atlantic Region
202/634-4126

South Atlantic Region
202/254-7494

Gulf Region
202/254-7546

Pacific Region
202/254-7100

Great Lakes Region
202/634-4124

*Indicates federally approved state programs.

Appendix B

Headquarters and Regional Offices of the U.S. Environmental Protection Agency*

*Summaries of revised water quality criteria documents for 64 toxic pollutants are available from EPA. These criteria are published pursuant to Section 304 (A)(1) of the Clean Water Act. The information is included in Part V of the Environmental Protection Agency's Water Quality Criteria Document: Availability *Federal Register*, Friday, November 28, 1980, Vol. 45, No. 231, FRL 1623-3.

Headquarters

Office of Environmental Review
Washington, DC 20460
202/755-8835

Office of Water and Waste Management
Clean Lakes Section
Washington, DC 20460
202/472-3400

Office of Water Program Operations
Washington, DC 20460
202/426-9404

Office of Water Planning and Standards
Washington, DC 20460
202/245-3145

Region 1

Rm. 2203 JFK Federal Bldg.
Boston, MA 02203
617/223-5131

Region 2

26 Federal Plaza, Rm. 1009
New York, NY 10007
212/264-4563

Region 3

Curtis Bldg., 6th and Walnut Streets
Philadelphia, PA 19106
215/597-7543

Region 4

345 Courtland Street, NE
Atlanta, GA 30308
404/881-4989

Region 5

230 South Dearborn Street
Chicago, IL 60604
312/353-2124

Region 6

1201 Elm Street
Dallas, TX 75270
214/767-2656

Region 7

324 E. 11th Street
Kansas City, MO 64106
816/374-5616

Region 8

1860 Lincoln Street
Denver, CO 80203
303/837-2721

Region 9

215 Fremont Street
San Francisco, CA 94105
415/556-7686

Region 10

1200 6th Avenue
Seattle, WA 98101
206/399-4011

Appendix C

Regional Offices of the U.S. Fish and Wildlife Service

Pacific Region

Regional Director
500 N.E. Multnomah Street
Portland, OR 97232
503/231-6118

Serves: California, Hawaii, Idaho, Nevada, Oregon, Washington

Southwest Region

Regional Director
Dennis Chaves Building
U.S. Court House
500 Gold Avenue, S.W.
Albuquerque, NM 87103
505/766-2321

Serves: Arizona, New Mexico, Oklahoma, Texas

North Central Region

Regional Director
Federal Building
Fort Snelling
Twin Cities, MN 55111
612/725-3563

Serves: Illinois, Indiana, Michigan, Minnesota, Ohio, Wisconsin

Southeast Region

Regional Director
Richard B. Russell Federal Building
75 Spring Street, S.W.
Atlanta, GA 30303
404/221-3554

Serves: Alabama, Arkansas, Florida, Georgia, Kentucky, Louisiana, Mississippi, North Carolina, South Carolina, Tennessee

Northeast Region

Regional Director
1 Gateway Center
Newton, MA 02158
617/965-5100

Serves: Connecticut, Delaware, Maine, Maryland, Massachusetts, New Hampshire, New Jersey, New York, Pennsylvania, Rhode Island, Vermont, Virginia, West Virginia

Alaska Region

Regional Director
1011 East Tudor Road
Anchorage, AK 99503
907/265-4808

Serves: Alaska

Denver Region

Regional Director
P.O. Box 25486
Denver Federal Center
Denver, CO 80225
303/234-2209

Serves: Colorado, Iowa, Kansas, Missouri, Montana, Nebraska, North Dakota, South Dakota, Utah, Wyoming

REGIONAL AND AREA OFFICE BOUNDARIES

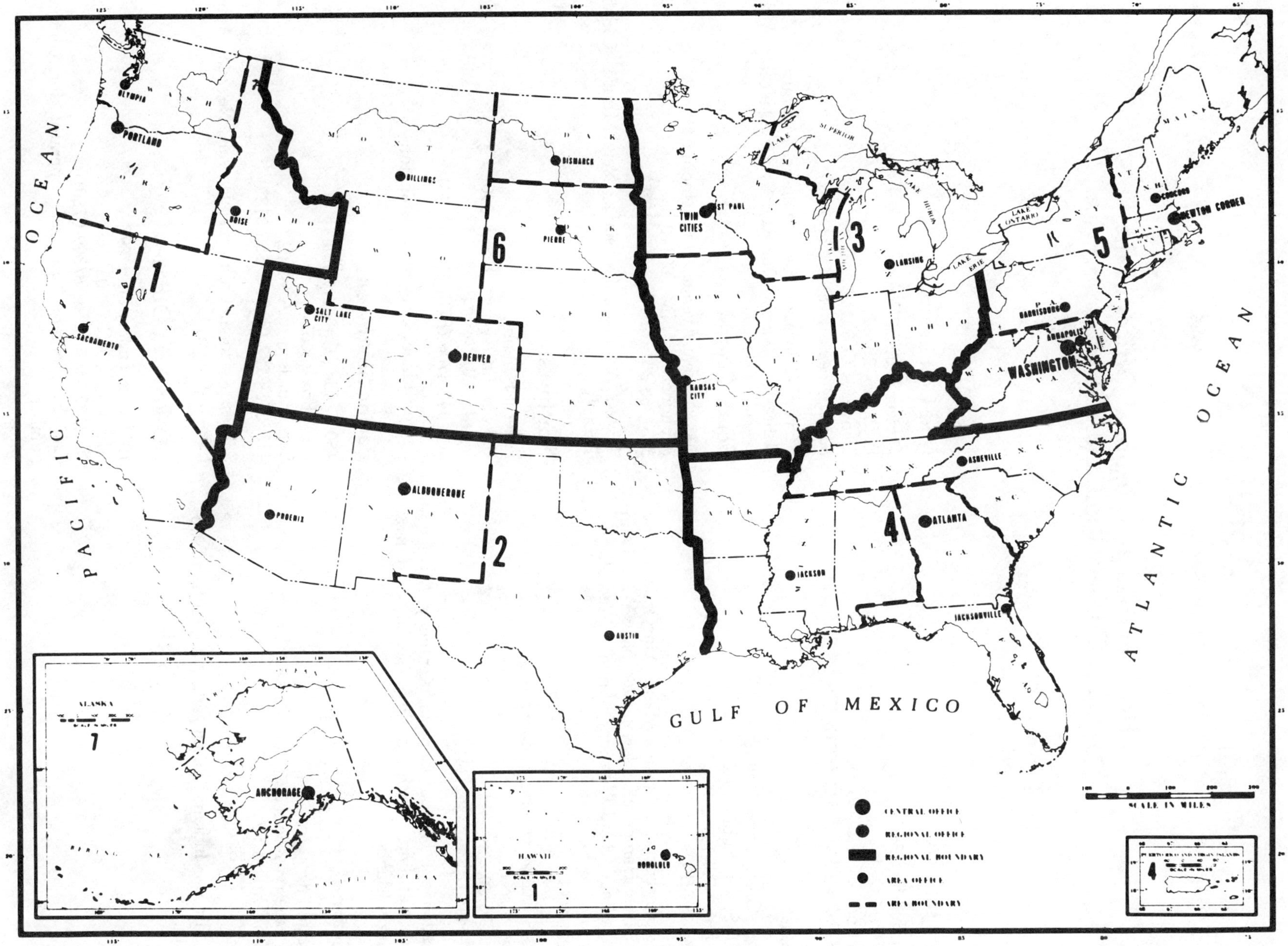

Fish and Wildlife Service Regions

Appendix D

Headquarters and Regional Offices of the Heritage Conservation and Recreation Service

Washington, DC

Pension Building
440 G Street, N.W.
Washington, DC 20243
202/343-5571 (FTS: 8/343-5571)

Northeast

William J. Green Federal Building
600 Arch Street, Room 9310
Philadelphia, PA 19106
215/597-7989 (FTS: 8/597-7990)

Serves: Connecticut, Delaware, Maine, Maryland, Massachusetts, New Hampshire, New Jersey, New York, Pennsylvania, Rhode Island, Vermont, Virginia, West Virginia

Southeast

Richard B. Russell Federal Building
75 Spring Street, S.W.
Atlanta, GA 30303
404/221-3445 (FTS: 8/242-3445)

Serves: Alabama, Florida, Georgia, Kentucky, Mississippi, North Carolina, South Carolina, Tennessee, Puerto Rico, Virgin Islands

Lake Central

Federal Building
200 East Liberty, Room 220
Ann Arbor, MI 48107
313/668-2023 (FTS: 8/378-2023)

Serves: Illinois, Indiana, Michigan, Minnesota, Ohio, Wisconsin

Mid-Continent

Denver Federal Center
P.O. Box 25387
Denver, CO 80225
303/234-6462 (FTS: 8/234-6462)

Serves: Colorado, Iowa, Kansas, Missouri, Montana, Nebraska, North Dakota, South Dakota, Wyoming

South Central

5000 Marble Avenue, N.E., Room 211
Albuquerque, NM 87110
505/766-3515 (FTS: 8/474-3515)

Serves: Arkansas, Louisiana, New Mexico, Oklahoma, Texas

Northwest

Federal Building
915 Second Avenue, Room 990
Seattle, WA 98174
206/442-4706 (FTS: 8/399-4706)

Serves: Idaho, Oregon, Washington

Pacific Southwest

450 Golden Gate Avenue
P.O. Box 36062
San Francisco, CA 94102
415/556-0182 (FTS: 8/556-0182)

Serves: Arizona, California, Hawaii, Nevada, American Samoa, Guam

Alaska Area Office

1011 East Tudor, Suite 297
Anchorage, AK 99503
907/277-1666 (FTS: 8/399-0150)
Call FTS number and ask operator for commercial number.

U.S. Department of Interior
HERITAGE CONSERVATION AND RECREATION SERVICE
Regional Offices and Regional Boundaries

Appendix E

Heritage Conservation and Recreation Service*
State Liaison Officers

ALABAMA

Commissioner
Department of Conservation and Natural Resources
Administrative Building
Montgomery, AL 36130
205/832-6361

ALASKA

Director
Alaska Division of Parks
619 Warehouse Drive, Suite 210
Anchorage, AK 99501
907/274-4676

AMERICAN SAMOA

Chairman of the Parks and Recreation Control Board of American Samoa
Pago Pago
Tutuila, AS 96799
653-5201 (via Overseas Operator)

ARIZONA

Director
Arizona Outdoor Recreation Coordinating Commission
1333 West Camelback Road, Suite 206
Phoenix, AZ 85013
602/255-5013 (FTS: 8/765-5013)

ARKANSAS

Director
Arkansas Dept. of Local Services
1 Capitol Mall
Little Rock, AR 72201
501/371-1211

CALIFORNIA

Director
Dept. of Parks and Recreation
P.O. Box 2390
Sacramento, CA 95811
8/916/445-2358

COLORADO

Director
Division of Parks and Outdoor Recreation
Department of Natural Resources
1313 Sherman Street, Room 618
Denver, CO 80203
303/839-3437

CONNECTICUT

Commissioner
Dept. of Environmental Protection
117 State Office Building
Hartford, CT 06115
203/566-2110

DELAWARE

Secretary
Dept. of Natural Resources and Environmental Control
Edward Tatnall Building
P.O. Box 1401
Dover, DE 19901
302/736-4403 (FTS: 8/487-6011)

DISTRICT OF COLUMBIA

Director
D.C. Recreation Department
3149 16th Street, N.W.
Washington, DC 20010
202/673-7665

FLORIDA

Director
Division of Recreation and Parks
Department of Natural Resources
3900 Commonwealth Blvd.
Tallahassee, FL 32303
904/488-6131

GEORGIA

Commissioner
State Dept. of Natural Resources
270 Washington Street, S.W.
Atlanta, GA 30334
404/656-3500

GUAM

Director
Dept. of Parks and Recreation
P.O. Box 2950
Agana, GU 96910
477-9620 (via Overseas Operator)

HAWAII

Chairman
Board of Land and Natural Resources
P.O. Box 621
Honolulu, HI 96809
808/548-6650

IDAHO

Director
Idaho Dept. of Parks and Recreation
Statehouse
Boise, ID 83720
208/334-2154

*See Preface.

ILLINOIS

Director
Department of Conservation
605 William G. Stratton Building
400 South Spring Street
Springfield, IL 62706
217/782-6302 (FTS: 8/956-6302)

INDIANA

Director
Department of Natural Resources
608 State Office Building
Indianapolis, IN 46204
317/633-6344 (FTS: 8/336-6344)

IOWA

Superintendent of Grants-in-Aid
Iowa Conservation Commission
Wallace State Office Building
Des Moines, IA 50319
515/281-5631

KANSAS

Director
Kansas Park and Resources Authority
503 Kansas Avenue
P.O. Box 977
Topeka, KS 66601
913/296-2281

KENTUCKY

Commissioner
Department of Local Government
Capitol Plaza Tower, 2nd Floor
Frankfort, KY 40601
502/564-2382 (FTS: 8/351-2382)

LOUISIANA

Asst. Secretary
Dept. of Culture, Recreation and
Tourism
Office of Program Development
P.O. Box 44247
Baton Rouge, LA 70804
504/925-3880 (FTS: 8/689-3880)

MAINE

Director
Bureau of Parks and Recreation
Department of Conservation
Statehouse—Sta. 19
Augusta, ME 04333
207/289-3821

MARYLAND

Deputy Secretary
Department of Natural Resources
Tawes State Office Building
Annapolis, MD 21401
301/269-3043

MASSACHUSETTS

Secretary
Department of Environmental Affairs
Leverett Saltonstall Building
100 Cambridge Street
Boston, MA 02202
617/727-9800

MICHIGAN

Deputy Director
Michigan Dept. of Natural Resources
Stevens T. Mason Building
P.O. Box 30028
Lansing, MI 48909
517/373-2682 (FTS: 8/253-2682)

MINNESOTA

Assistant Commissioner for Planning
Department of Natural Resources
Centennial Building, Box 51
St. Paul, MN 55155
612/296-6235 (FTS: 8/776-6235)

MISSISSIPPI

Director
Parks and Recreation
Department of Natural Resources
P.O. Box 10600
Jackson, MS 39209
601/961-5240

MISSOURI

Director
Department of Natural Resources
P.O. Box 176
Jefferson City, MO 65102
314/751-4422

MONTANA

Administrator
Parks Division
Montana Dept. of Fish, Wildlife
and Parks
1420 East 6th Avenue
Helena, MT 59601
406/449-3750

NEBRASKA

Director
Nebraska Game and Parks Commission
2200 N. 33rd Street
P.O. Box 30370
Lincoln, NE 68503
402/464-0641

NEVADA

Administrator
Nevada Division of State Parks
Capitol Complex
Carson City, NV 89710
702/885-4348

NEW HAMPSHIRE

Commissioner
Dept. of Resources and Economic
Development
P.O. Box 856
Concord, NH 03301
603/271-2411

NEW JERSEY

Deputy Commissioner
Dept. of Environmental Protection
P.O. Box 1390
Trenton, NJ 08625
609/292-0432

NEW MEXICO

Director
State Personnel Office
130 South Capital
Santa Fe, NM 87501
505/827-5201

NEW YORK

Commissioner
Office of Parks and Recreation
Agency Building #1
Empire State Plaza
Albany, NY 12238
518/474-0443

NORTH CAROLINA

Secretary
Dept. of Natural Resources and
Community Development
P.O. Box 27687
Raleigh, NC 27611
919/733-4984

NORTH DAKOTA

Director
North Dakota Dept. of Parks and Recreation
Box 139, R.R. #2
Mandan, ND 58554
701/663-3943

OHIO

Director
Department of Natural Resources
Fountain Square
Columbus, OH 43224
614/466-3770 (FTS: 8/942-3770)

OKLAHOMA

Executive Director
Oklahoma Tourism and Rec. Dept.
500 Will Rogers Memorial Building
Oklahoma City, OK 73105
405/521-2413

OREGON

State Parks Administrator
525 Trade Street, S.E.
Salem, OR 97310
503/378-6305

PENNSYLVANIA

Secretary
Dept. of Environmental Resources
P.O. Box 1467
Harrisburg, PA 17120
717/787-7160 (FTS: 8/637-7160)

PUERTO RICO

Administrator
Puerto Rico Recreational Dev. Co.
P.O. Box 2923
San Juan, PR 00903
809/725-1966

RHODE ISLAND

Director
Dept. of Environmental Management
Veterans Memorial Building
83 Park Street
Providence, RI 02903
401/277-2771

SOUTH CAROLINA

State Liaison Officer
Department of Parks, Recreation and Tourism
Suite 113, Edgar A. Brown Building
1205 Pendleton Street
Columbia, SC 29201
803/758-7705

SOUTH DAKOTA

Director
Division of Parks and Recreation
Department of Game, Fish and Parks
Sigurd Anderson Building
Pierre, SD 57501
605/773-3387 (FTS: 8/782-7000)

TENNESSEE

Commissioner
Department of Conservation
2611 West End Avenue
Nashville, TN 37203
615/741-1061

TEXAS

Associate Dean
University of Texas Law School
2500 Red River Road
Austin, TX 78705
512/471-3663

UTAH

Executive Director
Department of Natural Resources
231 East 400 South—Suite 400
Salt Lake City, UT 84102
801/533-5356

VERMONT

Secretary of the Agency of Environmental Conservation
79 River Street
Heritage 2 Building
Montpelier, VT 05602
802/828-3130

VIRGIN ISLANDS

Commissioner
Government of the Virgin Islands of the United States
Dept. of Conservation and Cultural Affairs
P.O. Box 4340
Charlotte Amalie
St. Thomas, VI 00801
809/774-3320

VIRGINIA

Director
Virginia Commission of Outdoor Recreation
8th & Franklin Streets
Old Federal Reserve Building
Richmond, VA 23219
804/786-2036

WASHINGTON

Administrator
Interagency Committee for Outdoor Recreation
4800 Capitol Blvd.
Tumwater, WA 98504
206/753-3610

WEST VIRGINIA

State Liaison Officer
Office of Economic and Community Development
Building 6—Room 522
Charleston, WV 25313
304/348-3361 (FTS: 8/885-3361)

WISCONSIN

Director
Office of Intergovernmental Programs
Department of Natural Resources
P.O. Box 7921
Madison, WI 53707
608/266-0836 (FTS: 8/366-0836)

WYOMING

Director
Wyoming Recreation Commission
604 East 25th Street
Cheyenne, WY 82002
307/777-7695

Appendix F

Regional Offices of the Water and Power Resources Service*

Pacific Northwest Region

Regional Director
P.O. Box 043, U.S. Court House
550 West Fort Street
Boise, ID 83724

Mid-Pacific Region

Regional Director
2800 Cottage Way
Sacramento, CA 95825

Lower Colorado Region

Regional Director
P.O. Box 427
Boulder City, NV 89005

Upper Colorado Region

Regional Director
P.O. Box 11568
Salt Lake City, UT 84111

Southwest Region

Regional Director
Commerce Building
714 South Tyler, Suite 201
Amarillo, TX 79101

Upper Missouri Region

Regional Director
P.O. Box 2553
Billings, MT 59103

Lower Missouri Region

Regional Director
Building 20, Denver Federal Center
Denver, CO 80225

Engineering and Research Center

Denver Federal Center, Building 67
P.O. Box 25007
Denver, CO 80225

*See Preface.

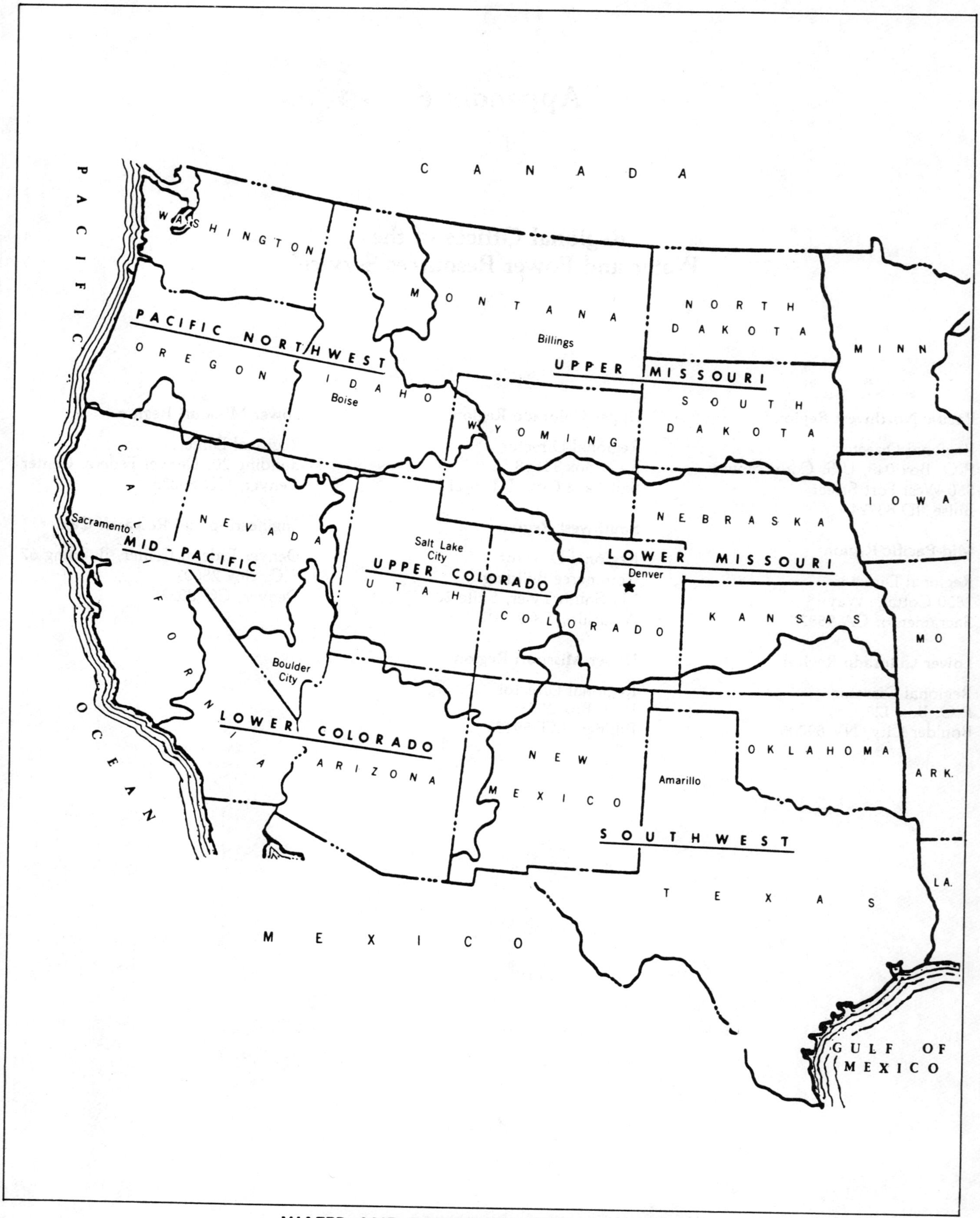

WATER AND POWER RESOURCES SERVICE
Regional Boundaries

Appendix G

Some Lake and Pond Inventory Considerations

(This is a partial listing of general considerations for inventory of water bodies. It is indended to alert the planner to some measurements that may be beneficial in planning for urban fishing. However, the planner should consult with state and local agencies for detailed information.)

General

Location	Drainage, state, county, geographic landmarks, latitude, longitude
Elevation	Feet above sea level, uppermost point to lowermost point
Climate	Mean temperature, rainfall
Origin/age	Glacial, impoundment, etc., approximate age

Physical

Size	Area in acres, volume in acre-feet
Depth profile	Depth contours
Fluctuation	High pool to low pool and frequency of change
Sources and discharges	Streams entering or leaving and annual discharge of each
Substrate	Substrate sizes, silt, sand, peat, rubble, etc., percent and area of each
Littoral zone size	Percent of total surface area and in acres
Shore line configuration	Length of shoreline compared to surface area
Stratification and turnover	Does seasonal stratification occur? Number of turnovers

Water Quality

Transparency/turbidity	Secchi disc or NTUs or JTUs seasonally, after rain and during dry period
Dissolved oxygen profile	Seasonally, during maximum temperature, in mg/l
Temperature profile	Seasonally, during stratification if it occurs
pH	Seasonally, at several locations and depths
Total dissolved solids	Seasonally, after rain if tributaries are present, at several locations and depths
Nutrients	Include total phosphorus, ammonia, nitrate, nitrite, total Kjeldahl nitrogen, seasonally, at several locations and depths
COD	Seasonally, at several locations and depths

Biological

Primary production	^{14}C fixation, chlorophyll *a* concentration, several locations
Phytoplankton	Dominant forms, phytoplankton, filamentous, densities (numbers/ml). Seasonally
Submergents	Species, categorize as rare, common, or abundant. Seasonally
Emergents	Same as previous
Floating-leafed	Same as previous
Zooplankton	Dominant forms, densities (numbers/m^3), seasonally
Fish	Species and numbers, general condition, tolerance, dominance, biomass
Waterfowl and other wildlife	Species and abundance, waterfowl, furbearers, small and big game, reptiles and amphibians, seasonally
Fecal coliform	Seasonally, several locations which reference potential contamination sources
BOD	Same as previous

Other

Land use practices	Land use classification, residential density, method of sewage disposal, proximity
Shoreline development	Docks, houses, industry, etc. Also bank stabilization or filling and dredging
Recreation	Public or private, fishing, boating, swimming, hunting, etc.
Point discharges	Location and type
Nuisance species	Algal blooms, rough fish, etc.
Special species	Rare and endangered plants or animals
Critical habitats	Spawning areas, migration routes, etc.
Intake points and volumes	Use, e.g., cooling water, municipal water supply, pump-storage, etc. Entrainment and impingement potential.
Subsurface geology	surficial soil infiltration rates, texture, bedrock fractures, chemical composition, depth, extent and type
hydrology	ground water flow rate, wells (location and pumping rate)
Other aspects of surrounding land uses	slope, surrounding bank vegetation, amount and concentration of nonpoint pollutants resulting from land uses and distances from water body

Appendix H

Some Wetland Inventory Considerations

(This is a partial listing of general considerations for inventory of water bodies. It is indended to alert the planner to some measurements that may be beneficial in planning for urban fishing. However, the planner should consult with state and local agencies for detailed information.)

General

Location	Drainage, state, county, geographic landmarks, latitude, longitude
Elevation	Feet above sea level, uppermost point to lowermost point
Climate	Mean temperature, rainfall
Type	Shrubswamp, bog, tidal marsh, etc.

Physical

Size	Surface area in acres
Configuration	Boundary length compared to total surface area
Hydrological location	Lakeside, streamside, estuary, deltaic, isolated, etc.
Open water areas	In acres and as a percent of total wetland area
Water source	If stream, discharge, drainage, etc.
Shoreline profile	Categorize as steep, medium, flat
Flow	Retention and rejuvenation of water, direction of movement, stagnation

Water Quality

Salinity	Report salinity from several locations in ppt. If water movement is significant measure seasonally
Turbidity	Seasonally in NTUs or JTUs
pH	Seasonally

Biological

Algae	Dominant forms, general abundance
Submergents	Species, categorize as rare, common, or abundant, seasonally
Emergents	Same as previous
Floating leafed	Same as previous
Cover type	Brush, trees, grasses, etc., relative proportion in percent and acreage
Vegetative interspersion	"Edge" lengths between cover types, open water, etc., compared to area
Fish	Species and abundance, general condition, tolerance, biomass, breeding use
Waterfowl and other wildlife	Species and abundance, waterfowl, furbearers, small and big game, reptiles and amphibians, seasonally

Other

Land use practices	Land use classification, residential density, method of sewage disposal, proximity
Filling or draining	Extent
Physical alterations/ hydrological barriers	Dikes and highways which may affect water movement
Recreation	Public or private, fishing, hunting, etc.
Point discharges	Location and type
Nuisance species	Algal blooms, fish, etc.

Special species	Rare and endangered plants and animals
Critical habitats	Spawning areas, migration routes, etc.
Subsurface	
geology	surficial soil infiltration rates, texture, bedrock fractures, chemical composition, depth, extent and type
hydrology	ground water flow rate, wells (location and pumping rate)
Other aspects of surrounding land uses	slope, surrounding bank vegetation, amount and concentration of nonpoint pollutants resulting from land uses and distances from water body

Appendix I

Some Stream and River Inventory Considerations

(This is a partial listing of general considerations for inventory of water bodies. It is indended to alert the planner to some measurements that may be beneficial in planning for urban fishing. However, the planner should consult with state and local agencies for detailed information.)

General

Location	Drainage, state, county, geographic landmarks, latitude, longitude
Elevation	Feet above sea level, uppermost point to lowermost point
Climate	Mean temperature, rainfall
Stream Order	Based on number of tributaries, from 1 to n

Physical

Basin Size	Square miles
Discharge, fluctuation	Mean annual, by month, cubic feet per second (cfs)
Floodplain size	Area inundated in square miles or acres by yearly high water and 25, 50 and 100 year floods
Channel width and cross-sectional area	Width in meters and cross section in square meters, average depth times width or by segments
Flooding history	25, 50, and 100 year flood discharges in cfs
Pool and riffle	Using several transects, the percent of each
Substrate	Boulder, rubble, cobble, gravel, sand, or silt, percent of each on several transects
Bank stability	Eroding or stable, extent of each
Gradient	Change in elevation per unit of stream length

Water Quality

Temperature	Average annual and seasonal or monthly in degrees Centigrade. Large rivers measured at several depths and locations
Dissolved oxygen	Consider backwaters and channel areas. In mg/l
Turbidity	During low flow and high flow (after rain) conditions, seasonally, in NTUs or JTUs
Specific conductance	During low flow and high flow conditions, seasonally, in μmhos/cm
Total dissolved solids	During low flow and high flow conditions, seasonally, in mg/l
Alkalinity	Same as previous
COD	During low flow and high flow conditions, seasonally
pH	Same as previous
Nutrients	Include total phosphorous, ammonia, nitrate, nitrite, total Kjeldahl nitrogen, during low flow and high flow, seasonally

Biological

Primary production	^{14}C fixation, chlorophyll *a* concentration
Algae	Dominant forms—filamentous, diatoms, periphyton, planktonic; indicate abundance of each form
Macrophytes	Types and locations, Abundance

Bank vegetation and shading	Trees, shrubs, herbs, grasses, percent of each, overhang extent
Macroinvertebrates	Number of taxa, abundance of each, general tolerance, of common taxa
Fish	Species and numbers, general condition, tolerance, dominance, biomass
Waterfowl and other wildlife	Species and abundance, waterfowl, furbearers, small and big game
Fecal coliform	Seasonally, several locations which reference point and non-point discharges
BOD	Same as previous

Other

Land use practices	Agriculture, urban, residential, natural, industrial; percent of each and square miles
Stream bank uses	Structures, docks, etc., or natural
Stream alterations	Channelization, damming, dredging, bank stabilization—reasons for each
Recreation	Sport or commercial fishing, boating, swimming, hunting, etc.
Point discharges	Location and type (municipal, industrial, type of industry, etc.)
Nuisance species	Algal blooms, mosquitoes, rough fish, etc.
Special species	Rare and endangered plants or animals
Intake points and volumes	Use—e.g., cooling water, irrigation, municipal water supply. Potential entrainment and impingement problems.
Critical habitats	Spawning areas, migration routes
Subsurface geology	surficial soil infiltration rates, texture, bedrock fractures, chemical composition, depth, extent and type
hydrology	ground water flow rate, wells (location and pumping rate)
Other aspects of surrounding land uses	slope, surrounding bank vegetation, amount and concentration of nonpoint pollutants resulting from land uses and distances from water body

Appendix J

U.S. EPA 1976 Quality Criteria for Water

Water Quality Characteristic or Pollutant	Criteria
Aesthetic Qualities	All waters free from substances attributable to wastewater or other discharges that: (1) settle to form objectionable deposits; (2) float as debris, scum, oil, or other matter to form nuisances; (3) produce objectionable color, odor, taste, or turbidity; (4) injure or are toxic or produce adverse physiological responses in humans, animals, or plants; and, (5) produce undesirable or nuisance aquatic life.
Alkalinity	20 mg/ℓ or more as $CaCO_3$ for freshwater aquatic life except where natural concentrations are less.
Ammonia	0.02 mg/ℓ (as un-ionized ammonia) for freshwater aquatic life.
Arsenic	50 μg/ℓ for domestic water supplies (health); 100 μg/ℓ for irrigation of crops.
Barium	1 mg/ℓ for domestic water supply (health).
Beryllium	11 μg/ℓ for the protection of aquatic life in soft fresh water; 1,100 μg/ℓ for the protection of aquatic life in hard fresh water; 100 μg/ℓ for continuous irrigation on all soils; except 500 μg/ℓ for irrigation on neutral to alkaline fine-textured soils.
Boron	750 μg/ℓ for long-term irrigation on sensitive crops.
Cadmium	10 μg/ℓ for domestic water supply (health). Aquatic Life:

Fresh Water

Soft Water	Hard Water	
0.4 μg/ℓ	1.2 μg/ℓ	for cladocerans and salmonid fishes;
4.0 μg/ℓ	12.0 μg/ℓ	for other, less sensitive, aquatic life.

Marine

5.0 μg/ℓ

Chlorine	Total residual chlorine: 2.0 μg/ℓ for salmonid fish; 10.0 μg/ℓ for other freshwater and marine organisms

Chromium	50 μg/ℓ for domestic water supply (health); 100 μg/ℓ for freshwater aquatic life.
Fecal Coliform Bacteria	**Bathing Waters**: Based on a minimum of not less than five samples taken over a 30-day period, the fecal coliform bacterial level should not exceed a log mean of 200 per 100 mℓ nor should more than 10 percent of the total samples taken during any 30-day period exceed 400 per 100 mℓ. **Shellfish Harvesting Waters**: Not to exceed a median fecal coliform bacterial concentration of 14 per 100 mℓ with not more than 10 percent of samples exceeding 43 per 100 mℓ for taking of shellfish.
Color	Waters shall be virtually free from substances producing objectionable color for aesthetic purposes; The source of supply should not exceed 75 color units on the platinum-cobalt scale for domestic water supplies; and Increased color (in combination with turbidity) should not reduce the depth of the compensation point for photosynthetic activity by more than 10 percent from the seasonally established norm for aquatic life.
Copper	1.0 mg/ℓ for domestic water supplies (welfare). For freshwater and marine aquatic life, 0.1 times a 96-hour LC_{50} as determined through nonaerated bioassay using a sensitive aquatic resident species.
Cyanide	5.0 μg/ℓ for freshwater and marine aquatic life and wildlife.
Gases. Total Dissolved	To protect freshwater and marine aquatic life, the total dissolved gas concentrations in water should not exceed 110 percent of the saturation value for gases at the existing atmospheric and hydrostatic pressures.
Iron	0.3 mg/ℓ for domestic water supplies (welfare); 1.0 mg/ℓ for freshwater aquatic life.
Lead	50 μg/ℓ for domestic water supply (health). 0.01 times the 96-hour LC_{50} value, using the receiving or comparable water as the diluent and soluble lead measurements (nonfilterable lead using an 0.45 micron filter), for sensitive freshwater resident species.
Manganese	50 μg/ℓ for domestic water supplies (welfare); 100 μg/ℓ for protection of consumers of marine mollusks.
Mercury	2.0 μg/ℓ for domestic water supply (health); 0.05 μg/ℓ for freshwater aquatic life and wildlife; 0.10 μg/ℓ for marine aquatic life.
Nickel	0.01 of the 96-hour LC_{50} for freshwater and marine aquatic life.
Nitrates, Nitrites	10 mg/ℓ nitrate nitrogen (N) for domestic water supply (health).
Oil and Grease	For domestic water supply: Virtually free from oil and grease, particularly from the tastes and odors that emanate from petroleum products. For aquatic life: 0.01 of the lowest continuous flow 96-hour LC_{50} to several important freshwater and marine species, each having a demonstrated high susceptibility to oils and petrochemicals. Levels of oils or petrochemicals in the sediment which cause deleterious effects to the biota should not be allowed.

Surface waters shall be virtually free from floating nonpetroleum oils of vegetable or animal origin, as well as petroleum derived oils.

Dissolved Oxygen

Aesthetics: Water should contain sufficient dissolved oxygen to maintain aerobic conditions in the water column and, except as affected by natural phenomena, at the sediment-water interface.

Freshwater aquatic life: A minimum concentration of dissolved oxygen to maintain good fish populations is 5.0 mg/ℓ. The criterion for salmonid spawning beds is a minimum of 5.0 mg/ℓ in the interstitial water of the gravel.

Aldrin-Dieldrin

0.003 μg/ℓ for freshwater and marine aquatic life.

The persistence, bioaccumulation potential and carcinogenicity of aldrin-dieldrin cautions human exposure to a minimum.

Chlordane

0.01 μg/ℓ for freshwater aquatic life;
0.004 μg/ℓ for marine aquatic life.

The persistence, bioaccumulation potential and carcinogenicity of chlordane cautions human exposure to a minimum.

Chlorophenoxy Herbicides

2, 4-D; 100 μg/ℓ for domestic water supply (health);
2, 4, 5-TP: 10 μg/ℓ for domestic water supply (health).

DDT

0.001 μg/ℓ for freshwater and marine aquatic life.

The persistence, bioaccumulation potential, and carcinogenicity of DDT cautions human exposure to a minimum.

Dementon

0.1 μg/ℓ for freshwater and marine aquatic life.

Endosulfan

0.003 μg/ℓ for freshwater aquatic life;
0.001 μg/ℓ for marine aquatic life.

Endrin

0.2 μg/ℓ for domestic water supply (health);
0.004 μg/ℓ for freshwater and marine aquatic life.

Guthion

0.01 μg/ℓ for freshwater and marine aquatic life.

Heptachlor

0.001 μg/ℓ for freshwater and marine aquatic life.

The persistence, bioaccumulation potential and carcinogenicity of heptachlor cautions human exposure to a minimum.

Lindane

4.0 μg/ℓ for domestic water supply (health);
0.01 μg/ℓ for freshwater aquatic life;
0.004 μg/ℓ for marine aquatic life.

Malathion

0.1 μg/ℓ for freshwater and marine aquatic life.

Methoxychlor

100 μg/ℓ for domestic water supply (health);
0.03 μg/ℓ for freshwater and marine aquatic life.

Mirex

0.001 μg/ℓ for freshwater and marine aquatic life.

Parathion

0.04 μg/ℓ for freshwater and marine aquatic life.

Toxaphene

5μg/ℓ for domestic water supply (health);
0.005 μg/ℓ for freshwater and marine aquatic life.

pH

Range: 5-9 Domestic water supplies (welfare);
6.5-9.0 Freshwater aquatic life;
6.5-8.5 Marine aquatic life.
(. . . but not more than 0.2 units outside of normally occurring range)

Phenol

1 μg/ℓ for domestic water supply (welfare), and to protect against fish flesh tainting.

Phosphorus

0.10 μg/ℓ yellow (elemental) phosphorus for marine or estuarine waters.

Phthalate Esters

3 μg/ℓ for freshwater aquatic life.

Polychlorinated Biphenyls

0.001 μg/ℓ for freshwater and marine aquatic life and for consumers thereof.
Every reasonable effort should be made to minimize human exposure.

Selenium

10 μg/ℓ for domestic water supply (health);
For marine and freshwater aquatic life; 0.01 of the 96-hour LC_{50} as determined through bioassay using a sensitive resident species.

Silver

50 μg/ℓ for domestic water supply (health);
For marine and freshwater aquatic life; 0.01 of the 96-hour LC_{50} as determined through bioassay using a sensitive resident species.

Solids (Dissolved) and Salinity

250 mg/ℓ for chlorides and sulfates in domestic water supplies (welfare).

Solids (Suspended, Settleable) and Turbidity

Freshwater fish and other aquatic life: Settleable and suspended solids should not reduce the depth of the compensation point for photosynthetic activity by more than 10 percent from the seasonably established norm for aquatic life.

Sulfide—Hydrogen Sulfide

2 μg/ℓ undissociated H_2S for fish and other aquatic life, fresh and marine water.

Tainting Substances

Materials should not be present in concentrations that individually or in combination produce undesirable flavors which are detectable by organoleptic tests performed on the edible portions of aquatic organisms.

Temperature

Freshwater Aquatic Life: For any time of year, there are two upper limiting temperatures for a location (based on the important sensitive species found there at that time):

(1) One limit consists of a maximum temperature for short exposures that is time dependent and is given by the species-specific equation:

$$\text{Temperature}_{(C^\circ)} = 1/b\,(\log_{10} \text{time}_{(min)} - a) - 2^\circ C$$

where: $\log_{10}$ = logarithm to base 10 (common logarithm)

a = intercept on the "y" or logarithmic axis of the line fitted to experimental data available from Appendix II-C, NAS, 1974 for some species.

b = slope of the line fitted to experimental data and available from Appendix II-C, NAS, 1974 for some species

and

(2) The second value is a limit on the weekly average temperature that:

a. In the cooler months (mid-October to mid-April in the north and December to February in the south) will protect against mortality of important species if the elevated plume temperature is suddenly dropped to the ambient temperature, with the limit the acclimation temperature −2°C when the lower lethal threshold temperature equals the ambient water temperature (in some regions this limitation may also be applicable in summer); or

b. In the warmer months (April through October in the north and March through November in the south) is determined by adding to the physiological optimum temperature (usually for growth) a factor calculated as one-third of the difference between the ultimate upper incipient lethal temperature and the optimum temperature for the most sensitive important species (and appropriate life state) that normally is found at that location and time.

c. During reproductive seasons (generally April through June and September through October in the north and March through May and October through November in the South) the limit is that temperature that meets site-specific requirements for successful migration, spawning, egg-incubation, fry-rearing, and other reproductive functions of important species. These local requirements should supersede all other requirements when they are applicable.

d. There is a site-specific limit that is found necessary to preserve normal species diversity or prevent appearance of nuisance organisms.

Source of Appendix J: Greenwood, D. R., G. L. Kingsbury, and J. G. Cleland. 1979. *A Handbook of Key Federal Regulations and Criteria for Multimedia Environmental Control.* U.S. Environmental Protection Agency, Washington, DC EPA-600/7-79-175, pp. 62-65.

Appendix K

Fish and Game Commissioners and Directors of the United States

Alabama Division of Game and Fish, 64 N. Union St., Montgomery 36104 (205, 832-6300)

Alaska Dept. of Fish and Game, Subport Bldg., Juneau 99801 (907, 465-4100)

Arizona Game and Fish Dept., 2222 W. Greenway Rd., Phoenix 85023 (602, 942-3000)

Arkansas Game and Fish Commission, Game and Fish Bldg., Little Rock 72201 (501, 371-1145)

California Dept. of Fish and Game, 1416 9th St., Sacramento 95814 (916, 445-3535)

Colorado Div. of Wildlife, 6060 Bdwy., Denver 80216 (303, 825-1192)

Connecticut Dept. of Environmental Protection, State Office Bldg., Hartford 06115 (203, 566-5460)

Delaware Div. of Fish and Wildlife, D St., Dover 19901 (302, 678-4431)

District of Columbia Metropolitan Police, 300 Indiana Ave., N.W., Washington 20001 (202, 626-2305)

Florida Game and Freshwater Fish Comm., 620 S. Meridian, Tallahassee 32304 (904, 488-1960)

Georgia State Game and Fish Div., Trinity-Washington Bldg., 220 Washington St., S.W., Atlanta 30334 (404, 656-3500)

Guam Dept. of Agriculture, Div. of Fish and Wildlife, Agana 96910 (722-6866)

Hawaii Div. of Fish and Game, 1179 Punchbowl St., Honolulu 96813 (548-4000)

Idaho Fish and Game Dept., 600 S. Walnut, Box 25, Boise 83707 (208, 384-3771)

Illinois Dept. of Conservation, State Office Bldg., Springfield 62706 (217, 782-6302)

Indiana Div. of Fish and Wildlife, 608 State Office Bldg., Indianapolis 46204 (317, 282-4020)

Iowa State Conservation Commission, State Office Bldg. 300 4th St., Des Moines 50319 (515, 281-5384)

Kansas Forestry, Fish and Game Commission, Box 1028, Pratt 67124 (316, 672-5911)

Kentucky Dept. of Fish and Wildlife Resources, Capitol Plaza Tower, Frankfort 40601 (502, 564-3400)

Louisiana Wildlife and Fisheries Commission, 400 Royal St., New Orleans 70130

Maine Dept. of Inland Fisheries and Game, 284 State St., Augusta 04330 (207, 289-3371)

Maryland Fish and Wildlife Administration, Natural Resources Bldg., Annapolis 21401 (301, 267 plus extension)

Massachusetts Div. of Fisheries and Wildlife, 100 Cambridge St., Boston 02202 (617, 727-3155)

Michigan Dept. of Natural Resources, Mason Bldg., Lansing 48926 (517, 373-1220)

Minnesota Dept. of Natural Resources, 301 Centennial Bldg., 658 Cedar St., St. Paul 55101 (612, 296-2894)

Mississippi Dept. of Conservation, P.O. Box 451, Jackson 39205 (601, 354-7333)

Missouri Dept. of Conservation, P.O. Box 180, Jefferson City 65101 (314, 751-4115)

Montana Department of Fish, Wildlife, and Parks, Helena 59601 (406, 499-3186)

Nebraska Game and Parks Commission, P.O. Box 30370, 2200 N. 33rd, Lincoln 68503 (402, 464-0641)

Nevada Dept. of Fish and Game, Box 10678, Reno 89510 (702, 784-6214)

New Hampshire Fish and Game Dept., 34 Bridge St., Concord 13301 (603, 271-3421)

New Jersey Div. of Fish, Game, and Shellfisheries, Box 1390, Trenton 08625 (609, 292-7348)

New Mexico, Natural Resources Department, Villagra Building, Santa Fe 87503 (505, 827-2923)

New York Div. of Fish and Wildlife, 50 Wolf Rd., Albany 12201 (518, 457-5690)

North Carolina Wildlife Resources Commission, 325 N. Salisbury St., Raleigh 27611 (919, 829-3391)

North Dakota State Game and Fish Dept., 2121 Lovett Ave., Bismarck 58501 (701, 224-2180)

Ohio Div. of Wildlife, Fountain Square, Columbus 43224 (614, 466-7313)

Oklahoma Dept. of Wildlife Conservation, 1801 N. Lincoln, P.O. Box 53465, Oklahoma City 73105 (405, 521-3851)

Oregon Fish and Wildlife Commission, Box 3503, Portland 97208 (503, 229-5551)

Pennsylvania Fish Commission, P.O. Box 1673, Harrisburg 17120 (717, 787-6593)

Pennsylvania Game Commission, P.O. Box 1567, Harrisburg 17120 (717, 787-3633)

Puerto Rico Ministry of Natural Resources, P.O. Box 5887, Puerta De Tierra, San Juan 00906

Rhode Island Dept. of Natural Resources, Div. of Fish and Wildlife, Washington County Government Center, Tower Hill Rd., Providence 02903 (401, 277-2784)

South Carolina Wildlife Resources Dept., Box 167, 1015 Main St., Columbia 29202 (803, 758-2561)

South Dakota Dept. of Game, Fish and Parks, State Office Bldg., Pierre 57501 (605, 224-3387)

Tennessee Wildlife Resources Agency, Box 40747, Ellington Agricultural Center, Nashville 37220 (615, 741-1431)

Texas Parks and Wildlife Dept., John H. Reagan Bldg., Austin 78701 (512, 475-8074)

Utah State Div. of Wildlife Resources, 1596 W. N. Temple, Salt Lake City 84116 (801, 328-5081)

Vermont Fish and Game Dept., Montpelier 05602 (802, 828-3371)

Virginia Commission of Game and Inland Fisheries, 4010 W. Broad St., Box 11104, Richmond 23230 (804, 770-4974)

Washington Dept. of Fisheries, 115 General Administration Bldg., Olympia 98504 (206, 753-6623)

Washington Dept. of Game, 600 N. Capitol Way, Olympia 98504 (206, 753-5700)

West Virginia Div. of Wildlife Resources, 1800 Washington St., East, Charleston 25305 (304, 348-2771)

Wisconsin Dept. of Natural Resources, Box 450, Madison 53701 (608, 266-2243)

Wyoming Game and Fish Dept., Box 1589, Cheyenne 82001 (307, 777-7631)

Source of Appendix K: *The Conservation Directory*, a publication of the National Wildlife Federation, 1412 16th Street, N.W., Washington, DC 20036.

Appendix L

Urban Waterfront Action Group
Directory of Participants

Urban waterfront revitalization is an issue of national importance and concern. Recognizing this, a task force—the Urban Waterfront Action Group—has formed to heighten awareness of opportunities, benefits and needs; address key issues; and explore cost-effective ways to respond to the needs of cities interested in revitalizing their waterfronts.

The following is a directory, provided as a public service by Partners for Livable Places, of Washington task force participants and the programs they represent. If you wish information about a specific program or referrals to a regional or field office, please contact the appropriate task force member.

Federal Agencies

United States Army Corps of Engineers

Regulatory Branch
Headquarters, Dept. of the Army
Office of the Chief of Engineers
Washington, DC 20314
Attn: DAEN-CWO-N
Phone: (202) 272-0200

Authorizes permits for activities in the waters of the United States.

Department of Commerce

Office of the Secretary
Regional Action Planning Commission
14th and Constitution Avenue, N.W.
Washington, DC 20230
Phone: (202) 377-4556

Provides assistance in gaining access to the planning, technical, and financial resources and services for local urban waterfront projects available from multi-state Regional Action Planning Commissions.

Office of the Secretary
Office of State and Local Government Assistance
14th and Constitution Avenue, N.W.
Washington, DC 20230
Room 5893
Phone: (202) 377-4556

The central point of contact for state and local government officials interested in gaining access to and coordinating Department of Commerce financial and non-financial resources and services for urban waterfront projects.

Economic Development Administration (EDA)
Office of Program Operations
14th and Constitution Avenue, N.W.
Washington, DC 20230
Room 7835
Phone: (202) 377-3027

CEDS is a new approach to helping communities plan and implement economic revitalization activity. In the CEDS process, communities develop investment strategies in which they identify local needs, set development priorities, and establish specific ways of addressing those priorities. Through this process, communities are able to better link investments by federal, state, and local agencies, and to stimulate increased private sector participation in economic development and job creation.

Office of Coastal Zone Management
Office of Policy and Evaluation
3300 Whitehaven Street, N.W.
Washington, DC 20235
Phone: (202) 634-4245

OCZM funds urban waterfront and harbor planning projects through state CZM grant programs. In addition, the office conducts research and offers advice and technical assistance pertaining to urban coastal issues.

Maritime Administration
Office of Port and Intermodal Development
14th and Constitution Avenue, N.W.
Washington, DC 20230
Phone: (202) 377-2424

Responsible for port promotion, planning and development on national, regional, state and local levels. The office sponsors: cost-shared comprehensive port planning studies which provide needs projections through the year 2000 and commercial port impact on urban waterfront development; computerized port facilities inventory which provides physical and operational characteristics of all U.S. public and private terminals; and technical assistance on local site development issues.

Department of Housing and Urban Development

Office of Environmental Quality
451 7th Street, S.W.
Washington, DC 20410
Phone: (202) 755-8909

Responsible for ensuring departmental compliance with the National Environmental Policy Act (NEPA), as well as overseeing HUD concerns relating to the physical, social, and economic environment.

Office of Policy Development and Research
451 7th Street, S.W.
Washington, DC 20410
Room 8146
Phone: (202) 755-7335 and 755-6450

Provides support services in the areas of long-range policy development, program evaluation, and research for the program offices within the Department.

Office of Community Planning and Development
451 7th Street, S.W.
Washington, DC 20410
Room 7224
Phone: (202) 755-6240

Provides funding to local governments through the Community Development Block Grant, the Urban Development Action Grant, Section 312 Rehabilitation. Also provides Section 701 Comprehensive Planning Assistance and various Innovative Grants from the Secretary's Discretionary Fund.

Department of the Interior

Heritage Conservation and Recreation Service
Technical Preservation Services
(Maritime Heritage Program)
440 G Street, N.W.
Washington, DC 20243
Room 230A
Phone: (202) 343-7217
343-6384
343-4256
343-6295

This office administers a five million dollar maritime preservation grants-in-aid program to the states and to the National Trust for Historic Preservation project. Categories include urban waterfront development, planning, engineering and architectural services, historic vessels, and maritime educational programs.

Heritage Conservation and Recreation Service
Water Resources Section
440 G Street, N.W.
Washington, DC 20243
Room 312
Phone: (202) 343-5571
343-7801
343-4693

The Water Resources Section conducts studies, reviews projects and plans, and provides a variety of technical assistance to communities interested in urban waterfront revitalization.

National Park Service
Office of Park Planning and Environmental Quality
18th and C Street, N.W.
Washington, DC 20240
Room 3013
Phone: (202) 343-5625

This office is interested in the relationship of urban waterfront projects to existing parks or proposed park sites.

Department of Transportation

Office of the Secretary
Office of Intergovernmental Affairs, I-23
400 7th Street, S.W.
Washington, DC 20590
Room 10405
Phone: (202) 426-0163

Coordinates the development of transportation systems as they impact upon urban and rural areas.

Environmental Protection Agency

Clean Lakes Program
Criteria and Standards Division (WH-585)
401 M Street, S.W.
Washington, DC 20460
Room 2812M
Phone: (202) 472-3400

Provides technical and financial assistance to restore and protect the water quality and usability of publicly owned freshwater lakes. EPA has recently announced an urban lakes initiative which will provide an additional focus for the program on urban lakes.

National Workforce Development Staff
401 M Street, S.W.
Washington, DC 20460
Phone: (202) 755-8835

Coordinates the EPA Urban Work Group, which was established by EPA to deal with urban-specific problems and to carry out EPA's portion of the current administration's urban policy.

Federal Insurance Administration, Federal Emergency Management Agency

Program Analysis and Evaluation Division
451 7th Street, S.W.
Washington, DC 20410
Room 5264
Phone: (202) 426-1891

The National Flood Insurance Program, administered by the Federal Insurance Administration, enables persons to purchase insurance against losses from damage or destruction of real or personal property caused by floods or flood-caused erosion, and to promote wise floodplain management practices in the nation's flood-prone areas.

National Endowment for the Arts

Design Arts
2401 E Street, N.W.
Washington, DC 20506
Phone: (202) 634-4286

Small grants (maximum of $30,000) are awarded to non-profit organizations, including local governments, for planning and design work. The grants can be used for the conceptualization necessary to initiate waterfront revitalization. The grant awards must be matched 50% by the organization.

National Endowment for the Humanities

Special Assistant to the Chairman
806 15th Street, N.W.
Mail Stop 302
Washington, DC 20506
Phone: (202) 724-0297

The Endowment supports waterfront projects in the humanities including historic documentation and preservation, media projects, exhibits, research, and issue discussion at the national and state levels.

Non-Federal Organizations

National Trust for Historic Preservation

Maritime Preservation Office
1785 Massachusetts Avenue, N.W.
Washington, DC 20036
Phone: (202) 673-4127

This office encourages and assists public agencies and private organizations to include maritime heritage focus in urban waterfront revitalization projects through adaptive use of historic maritime resources ashore and afloat. Provides technical and advisory services and financial assistance through maritime heritage preservation grants program.

United States Conference of Mayors

Railroad Land Revitalization Program
1620 I Street, N.W.
Washington, DC 20006
Suite 510
Phone: (202) 293-6910

This Department of Commerce-funded program provides technical assistance to cities and railroads to relocate railroad facilities and to help overcome barriers to the redevelopment of railroad properties in cities. Many ports and waterfront areas are encumbered by rail facilities which act as barriers to reuse. This program can advise public and private agencies as to how these barriers can be overcome.

National League of Cities

Urban Environmental Design Project
1620 I Street, N.W., 2nd floor
Washington, DC 20006
Phone: (202) 293-6795

The National League of Cities represents over 800 cities directly and over 15,000 through their state municipal leagues in Washington. The League provides its members with technical assistance in a variety of subject areas, one of which is urban environmental design. NLC's Urban Environmental Design Project assists cities to improve the design management and quality in their development processes.

Partners for Livable Places

2120 P Street, N.W.
Washington, DC 20037
Phone: (202) 223-5867

National non-profit coalition which works to improve the quality of life in our communities through the conservation and sensitive development of the physical environment. 'Partners' encourages public/private partnerships such as that represented by UWAC. It also operates an information clearinghouse and referral service on the built environment, which includes waterfronts.

Appendix M

Regional Offices of the National Marine Fisheries Service

Alaska Region

(State of Alaska and the Pribilof Islands.) Director, Alaska Region, National Marine Fisheries Service, NOAA, P.O. Box 1668, Juneau, AK 99802

Northeast Region

(Connecticut, Delaware, Illinois, Indiana, Maine, Maryland, Massachusetts, Michigan, Minnesota, New Hampshire, New Jersey, New York, Ohio, Pennsylvania, Rhode Island, Vermont, Virginia, West Virginia, and Wisconsin.) Director, Northeast Region, National Marine Fisheries Service, NOAA, 14 Elm Street, Federal Bldg., Gloucester, MA 01930

Northwest Region

(Colorado, Idaho, Montana, North Dakota, Oregon, South Dakota, Utah, Washington, and Wyoming.) Director, Northwest Region, National Marine Fisheries Service, NOAA, 1700 Westlake Ave., North, Seattle, WA 98109

Southeast Region

(Alabama, Arkansas, Florida, Georgia, Iowa, Kansas, Kentucky, Louisiana, Mississippi, Missouri, Nebraska, New Mexico, North Carolina, Oklahoma, South Carolina, Tennessee, Texas.) Director, Southeast Region, National Marine Fisheries Service, NOAA, 9450 Koger Blvd., St. Petersburg, FL 33702

Southwest Region

(Arizona, California, and Nevada.) Director, Southwest Region, National Marine Fisheries Service, NOAA, 300 S. Ferry Street, Terminal Island, CA 90731

Pittsburgh Press photo by Ross Catanza

Figure 39. *Outings on the waterfront provide opportunities for food, fishing, and fun. A fishing contest highlighted this picnic on the Pittsburgh waterfront.*

BIBLIOGRAPHIC DATA SHEET	1. Report No. FWS/OBA-80/35	2.	3. Recipient's Accession No.

4. Title and Subtitle
Planning for Urban Fishing and Waterfront Recreation

5. Report Date July 1981

6.

7. Author(s)
D. L. Leedy, T. M. Franklin, R. M. Maestro

8. Performing Organization Rept. No.

9. Performing Organization Name and Address
Urban Wildlife Research Center, Inc.
10921 Trotting Ridge Way
Columbia, MD 21044

10. Project/Task/Work Unit No.

11. Contract/Grant No. 14-16-0009-79-1014

12. Sponsoring Organization Name and Address
Fish and Wildlife Service (DOI); Heritage Conservation and Recreation Service (NPS/DOI); Water and Power Resources Service (BuRec); Water Resources Institute (COE); Clean Lakes Program (EPA); Sport Fishing Institute; Academy of Natural Sciences of Philadelphia (see reverse for addresses)

13. Type of Report & Period Covered final report

14.

15. Supplementary Notes

16. Abstracts
This document is intended to help planners and developers enhance fishing and waterfront recreation in urban areas by preserving existing high-quality aquatic areas, restoring degraded areas, and creating new areas where appropriate. Background information about environmental, fish and wildlife values, the nature of aquatic ecosystems, urbanization effects on these systems, planning and management implications and the importance of aquatic resource considerations in urban-suburban planning are discussed. Also discussed is planning for urban fishing and related waterfront recreation, including a review of literature and guidelines for preserving and enhancing urban aquatic resources and water-related recreational opportunities. Zoning, liability, safety, and other considerations are treated also. The document also includes detailed guidance on steps to be taken by planners to incorporate considerations for aquatic resources into the planning and decisionmaking process to ensure that healthy water bodies will be retained, degraded waters will be rehabilitated, and new water bodies will be created for specific or multi-

17. Key Words and Document Analysis. 17a. Descriptors

municipal planning, city planning, urban planning, recreation

17b. Identifiers/Open-Ended Terms
regional planning, fishing, fisheries, aquatic biology, recreational facilities, fish, environmental planning

17c. COSATI Field/Group 48B, G; 57H; 68D; 70F; 91A, H, L

18. Availability Statement	19. Security Class (This Report) UNCLASSIFIED	21. No. of Pages viii+106
Unlimited	20. Security Class (This Page) UNCLASSIFIED	22. Price

FORM NTIS-35 (REV. 10-73) ENDORSED BY ANSI AND UNESCO. USCOMM-DC 8265-P74

BIBLIOGRAPHIC DATA SHEET (continued)   FWS/OBS-80/35

12. Addresses of sponsoring organizations:

U.S. Fish and Wildlife Service
Department of the Interior
Eastern Energy and Land Use Team
Route 3 Box 44
Kearneysville, WV 25430

Water Resources Institute
U.S. Army Corps of Engineers
Washington, DC

Clean Lakes Program
U.S. Environmental Protection Agency
Washington, DC 20240

Academy of Natural Sciences of Philadelphia
Philadelphia, PA 19103

Heritage Conservation and Recreation Service
U.S. Department of the Interior
Washington, DC 20240

Water and Power Resources Service
Bureau of Reclamation
U.S. Department of the Interior
Washington, DC

Sport Fishing Institute
608 13th Street, NW
Washington, DC

16. Abstract, continued:

purpose uses. A chapter is included on site planning and on regional/municipal planning. Additional sources of technical information, financial assistance, funding opportunities and their legal requirements, and recommended readings are included. The appendices include lists of relevant Federal, regional and State agencies, and tables useful in inventorying aquatic resources.